DEPARTMENT OF TRANSPORT

Vehicle Testing

THE M.O.T. TESTER'S MANUAL

including

Explanatory Notes on the Statutory Provisions & Regulations for Testing of Motor Vehicles (except Motor Cycles) and Light Goods Vehicles, under Section 43 of the Road Traffic Act 1972.

LONDON: HER MAJESTY'S STATIONERY OFFICE

First published 1976
Ninth impression 1988

ISBN 0 11 550406 0

Preface

The purpose of this Manual, issued with the approval of the Secretary of State for Transport, is to serve as a working guide for those who carry out the statutory test on passenger cars and light goods vehicles. The owner of a vehicle should also find the Manual useful in that it details the inspections to which his vehicle should be subjected and the principal reasons why it may not be issued with a test certificate.
The contents of the Manual should not be regarded as a substitute for the statutory provisions and regulations.

Contents

Introduction
Notes on the use of the Manual

1 This Manual is a guide to the inspection procedures to be adopted for the statutory test on passenger and light goods vehicles having three or more wheels, except motor cycles with side-cars attached. It deals in detail with the inspection procedures.

2 The order in which the sections are set out in the Manual is not necessarily that in which the inspection of the items is to be carried out in practice. This is because some of the inspections are conducted concurrently by the tester and an assistant and there are instances where a part of one inspection has a bearing on the inspection of another part. The order in which the inspections are carried out will, therefore, depend on the particular vehicle concerned and the layout and type of the equipment in the testing station. The two diagrams in Appendix A show typical inspection routines which will aid the tester in making a thorough inspection of a vehicle, although this routine may have to be varied to suit the different types of inspection equipment in use.

3 The Manual is written on the assumption that the inspection will be carried out by a tester, who has been nominated for this purpose, together with an assistant working under his direction. In appropriate cases the person submitting the vehicle may carry out the functions of the assistant, provided the nominated tester is satisfied that he is competent to do so. For example, the person submitting the vehicle will normally be capable of switching on the lighting equipment, operating the steering or the brakes when the nominated tester requests him to do so. It is, however, necessary to emphasize that only a nominated tester is empowered to make a decision about the results of the inspection of a particular item.

4 The "Method of Inspection" column details the way in which the inspection of items on the vehicle is to be carried out and the equipment to be used. When carrying out each inspection, particular attention should be paid to the information given in the "Notes" column, since this gives valuable guidance on the conduct of the test and the scope of the various inspections.

5 The "Principal Reasons for Failure" column gives guidance on the types of defects which result in the vehicle failing to meet the statutory requirements. Having regard to the varying types of construction and the many different models of vehicle subject to test (or likely to be subject to test in future), it is not possible to say with certainty that every defect which might occur on a vehicle has been listed.

6 Because it is not practicable to lay down limits of

wear and tolerances for all types of components of different models of vehicle, a tester is expected to use his experience and judgement in making his assessment of the condition of a component. The main criteria he should use when making such an assessment are:

(i) whether the component has or has not reached the stage where it is obviously likely to affect adversely the roadworthiness of the vehicle;

(ii) whether or not the components has clearly reached the stage when replacement, repair or adjustment is necessary and

(iii) whether the condition of the component appears to contravene the requirements of the law.

7 If a defect in a testable item is found at an inspection, the nature of the defect must be clearly stated in Section A of the Inspection Report (form VT30).

8 If, during the course of an inspection, a defect is seen in a component (whether it is a testable item or not) and the defect is such that in the opinion of the tester the vehicle is likely to cause danger to any person or damage to the vehicle or any other property when driven on the road, details of this defect must be stated in Section D of the Inspection Report (VT30).

9 Appendix A illustrates a typical routine for vehicle inspection.

Appendix B shows the Inspection Report (VT30) which incorporates notification of failure.

Appendix C gives guidance on the assessment of corrosion on a vehicle.

Appendix D gives information on the prescribed Statutory Requirements for Classes III, IV and V vehicles.

10 The statutory test does not specifically include a road test of the vehicle, except in those cases where the testing station has not been required to install a roller brake testing machine, or for those vehicles which cannot have the performance of their brakes tested on a roller brake testing machine. A tester may, however, carry out a road test if he so wishes.

11 It is a requirement that the statutory test must only be conducted on the equipment which has been designated as acceptable for the test, and that the designated equipment must always be used for the test.

12 A tester need not start a test in any of the following circumstances:

(i) when it is necessary, for the proper conduct of the test, to know the age of the vehicle and the owner is unable to provide satisfactory evidence of this;

(ii) the vehicle is not capable of being moved under its own power during the test;

(iii) the vehicle is in such a dirty condition that an inspection cannot reasonably be carried out;

(iv) the load or other items are insecure;

(v) the vehicle is of such a size or weight that it cannot be tested on the approved facilities.

13 A tester need not complete an inspection when he discovers some item of an advisory or more serious nature, during the inspection which, in his opinion, makes it essential to discontinue the inspection on the grounds of safety. In such a case the reasons should be stated in Section C of the Inspection Report (VT30).

Section I

Lighting Equipment, Stop Lamps, Reflectors, and Direction Indicators

Method of inspection	Principal reasons for failure	Notes
1 FUNCTION OF OBLIGATORY FRONT AND REAR LAMPS		
Obligatory front lamps are the two "sidelamps", which are required by regulation, to be provided on a vehicle showing a WHITE light, which must be visible from a reasonable distance, to the front of the vehicle.	1 There are not two unobscured sidelamps with diffused lenses, symmetrically placed, showing a white light to the front, which is visible from a reasonable distance. (See Note 3).	1 *This inspection applies generally to all vehicles, except those not used on roads during the hours of darkness, which either have no lamps at all, or have lamps which are permanently disconnected or masked. (See Appendix D Section B).*
Obligatory rear lamps are the two lamps, which are required by regulation, to be provided on a vehicle showing a RED light, which must be visible from a reasonable distance, to the rear of the vehicle.	2 There are not two unobscured rear lamps with diffused lenses, symmetrically placed, showing a red light to the rear, which is visible from a reasonable distance. (See Note 3).	2 ***The obligatory front lamps may be incorporated with the headlamps (which may be yellow) but must be switched separately.***
With the side and rear lamps switched on, see that they each show a light of the correct colour and of sufficient intensity to enable them to be seen from a reasonable distance.	3 **A lamp showing a light other than white to the front. (See Note 2).**	3 *The measurement of the precise position of obligatory front and rear lamps is not part of the inspection, but nevertheless it is necessary to check visually that the lamps are located at approximately the same height and the same distance inboard from the side of the vehicle.*
	4 A lamp showing a light other than red to the rear.	
	5 A sidelamp or rear lamp with a damaged or missing lens.	
	6 Sidelamps obviously not located at the same height. (See Note 3).	
	7 Rear lamps obviously not located at the same height. (See Note 3).	

Method of inspection	Principal reasons for failure	Notes
	8 An obligatory front or rear lamp which does not illuminate immediately it is switched on because of a poor electrical connection.	

2 FUNCTION OF OBLIGATORY HEADLAMPS

Method of inspection	Principal reasons for failure	Notes
Obligatory headlamps are lamps with centres located between 60 cm (24") and 106 cm (42") above ground level, which must be provided on a vehicle to illuminate the road in front of it during the hours of darkness. They may be white or yellow in colour. Any additional lamps, such as fog lamps, are not obligatory headlamps.	1 There are not at least two unobscured white or two unobscured yellow lights placed approximately symmetrically, one on each side of the vehicle, capable of illuminating the road in front of the vehicle. (See Note 2).	*1 This inspection applies generally to all vehicles, except those not used during the hours of darkness, which either have no lights at all or have lights which are permanently disconnected or masked. (See Appendix D Section B).*
2.1 With the obligatory headlamps switched on main and dipped beam in turn, see that they each show as selected a light both of the same colour (white or yellow) of sufficient intensity to illuminate the road in front of the vehicle.	2 An obligatory headlamp which does not illuminate on dipped beam and on main beam as selected.	*2 The measurement of the precise position of obligatory headlamps is not part of the inspection, but nevertheless it is necessary to check visually that the lamps are located at approximately the same height and the same distance inboard from the side of the vehicle.*
For the inspection of headlamp aim and the correct functioning of the dipping mechanism see subsection 6.	3 An obligatory headlamp which does not illuminate on both main and dipped beam, immediately it is switched on because of a poor electrical connection.	*3 The inspection does not include the functioning or position of fog or auxiliary lamps.*
	4 Two obligatory headlamps showing different colours. For example one yellow and one white. (See Note 4).	*4 Each lamp in a matched pair of obligatory headlamps must emit light of the same colour. In the case of a four headlamp system the outer pair may emit light of one colour and the inner pair a different colour. There is no requirement for each lamp in a pair of non obligatory headlamps to emit light of the same colour.*
	5 The beams from obligatory headlamps are not deflected downwards when the dipping mechanism is operated.	
	6 An obligatory headlamp with a lens missing, damaged or located incorrectly in the lamp housing.	
	7 A headlamp showing a light well below the intensity required to illuminate the road in front of the vehicle.	

Method of inspection	Principal reasons for failure	Notes

3 FUNCTION OF STOP LAMPS

Method of inspection	Principal reasons for failure	Notes
A stop lamp is a lamp fitted to a motor vehicle for the purpose of warning other road users when the lamp is lit, that the brakes of the motor vehicle are being applied. 3.1 With the ignition switched on and the service brake (footbrake) applied, observe the functioning of the stop lamp(s).	A For vehicles first used BEFORE 1 January 1971 : 1 The red stop lamp(s) at the rear of the vehicle is missing, or does not illuminate when the service brake is applied. 2 The stop lamp does not remain steadily illuminated when the service brake is applied. 3 The stop lamp remains illuminated, even when the service brake is not applied. 4 A stop lamp with a damaged or missing diffused lens. B For vehicles first used on or AFTER 1 January 1971 : 1 There are not two unobscured red stop lamps, placed approximately symmetrically one on each side of the vehicle, which illuminate when the service brake is applied. (See Note 3). 2 The light from either of the stop lamps does not remain steadily illuminated when the service brake is applied. 3 A stop lamp remains illuminated, even when the service brake is not applied. 4 A stop lamp with a damaged or missing diffused lens.	1 *Vehicles first used BEFORE 1 January 1971 need only have one stop lamp positioned from the centre to the right-hand side (off-side) of the vehicle at the rear, but they are permitted to have two stop lamps, one on each side, at the rear. If such vehicles do have two stop lamps they should be treated as vehicles first used on or after 1 January 1971. Vehicles first used on or after 1 January 1971 must have two stop lamps, one on each side of the vehicle, at the rear. Some vehicles, however, do not have to be fitted with stop lamps. (See Appendix D Section C1).* 2 *On vehicles first used BEFORE 1 September 1965, a stop lamp may be combined with a rear direction indicator.* 3 *The measurement of the precise position of stop lamps is not part of the inspection, but where two stop lamps are fitted, it is necessary to check visually that each lamp is located at approximately the same height and the same distance inboard from the side of the vehicle.*

4 OBLIGATORY REAR REFLECTORS

Method of inspection	Principal reasons for failure	Notes
4.1 Examine the condition and fixing of reflectors on the rear of the vehicle, and see that they are red in colour.	1 There are not two unobscured red reflex reflectors fitted squarely, and approximately symmetrically, one on each side of the vehicle at the rear. (See Note 2).	1 *This inspection applies to all vehicles.* 2 *The measurement of the precise position of rear reflectors is not part of the inspection, but nevertheless it is necessary to check visually that*

Method of inspection	Principal reasons for failure	Notes
		all reflectors are located at approximately the same height and the same distance inboard from the side of the vehicle.
	2 A part of the reflecting area of a reflector is missing.	
	3 A reflector is not securely fixed to the vehicle.	*3 Any additional reflectors fitted to a vehicle, other than reflectors required by Regulation, are not included in the statutory inspection.*
		4 The inspection does not include a check that reflectors have the appropriate approval mark. Reflecting tape may not be regarded as a substitute for a rear reflector.

5 FUNCTION OF DIRECTION INDICATORS

Method of inspection	Principal reasons for failure	Notes
5.1 Semaphore Type With the ignition switched on and the direction indicators operated in turn on each side, see whether the semaphore arms are moving from the side of the vehicle correctly and are not sticking. See that each arm is illuminated, amber in colour, on both front and rear faces. Whilst the indicators are operating ,see that the "tell tale" is recording the operation of the indicators correctly, or alternatively that each direction indicator, when in operation, is readily visible from the driving seat.	1 A semaphore arm which does not move out to a horizontal position without sticking. 2 A semaphore arm which does not move back to the "hidden" position without sticking. 3 For vehicles with electric lighting, a semaphore arm which does not illuminate when in operation. 4 For vehicles with electric lighting, a semaphore arm which illuminates, but shows a light other than amber to the front or rear. 5 The "tell tale" does not indicate the operation of the indicators correctly, or the indicators when in operation, are not readily visible from the driving seat.	*1 A vehicle which is not equipped with electric lighting need not have semaphore type direction indicators which illuminate.* *2 Vehicles first used BEFORE 1 January 1936, and vehicles which cannot exceed 24 kph (15 mph) under their own power, need not be equipped with direction indicators. If, however, vehicles are fitted with direction indicators, they must meet the requirements of this inspection. (See Appendix D Section D).* *3 The position and angle of visibility of direction indicators are not part of the inspection.* *4 The correct colour for semaphore and flashing direction indicators depends on the date of first registration of the vehicle.*
5.2 Flashing Type With the ignition switched on and the direction indicators operated in turn on each side, see that they are flashing at approximately the correct rate. That is, between one and two flashes per second.	1 A direction indicator not operating, or not flashing at approximately the correct rate. 2 A direction indicator showing to the front a light other than white or amber.	*5 Any additional indicators fitted to a vehicle are not included in the inspection.*

Method of Inspection	Principal reasons for failure	Notes
See that the direction indicators at the front are either white or amber in colour. See that direction indicators at the rear are either red or amber in colour. Whilst operating the indicators see that the "tell tale" is recording the operation of the indicators correctly, or alternatively that each direction indicator is readily visible from the driving seat.	3 A direction indicator showing to the rear a light other than amber or red. 4 The "tell tale" does not indicate the operation of the direction indicators correctly, or if no "tell tale" is fitted the indicators are not visible from the driving seat. 5 A direction indicator with a lens missing or damaged, or with a lens which does not diffuse the light, or one which is obscured.	6 *In some cases, the rate of flashing of direction indicators may be affected by the condition of the vehicle's battery. It may, therefore, be necessary to run the engine whilst checking the rate of flashing of direction indicators.*

6 AIM OF HEADLAMPS

6.1 Headlamp Types

Headlamps fall into three broad types depending on whether they are designed for setting on main (driving) beam or on dipped (passing) beam. When conducting the statutory test it is necessary to know the headlamp type in order to establish the method of test to ensure that it is capable of emitting a beam which does not dazzle. Information on how to determine the type of headlamp is given on Pages I/7, I/8, I/9 and I/10.

6.2 When checking the aim of headlamps proceed as follows:

Locate the vehicle on the area designated as the "standing area" for the headlamp aim test.

With a person sitting in the driving seat, switch the headlamps on so that they emit the appropriate beam on which the headlamp aim is to be checked. See pages I/7, I/8, I/9 and I/10.

Align the headlamp aim equipment with the longitudinal axis of the vehicle and locate the centre of the collecting lens with the centre

Principal reasons for failure

1 British American Type Headlamp
Checked on main (driving) beam (Page I/7).

(i) The centre of the area of maximum intensity is above the horizontal line of the aiming screen, (Diagram 3).

(ii) The centre of the area of maximum intensity is more than 1.2 to the offside of the vertical line of the aiming screen, (Diagram 3) (See Note 5 on maximum illumination).

(iii) When dipped the brightest part of the image does not move downwards or downwards and to the nearside. That is the lamp or bulb is for a right hand rule of the road vehicle.

2 British American Type Headlamp
Checked on dipped (passing) beam (Page I/8).

(i) The upper edge of the maximum intensity area (hot spot) is more than 0.25° above the horizontal line of the aiming screen (Diagram 6).

(ii) The offside edge of the maximum intensity area (hot spot) is to the offside of the vertical line of the aiming screen (Diagram 6). (See Note 5 on maximum illumination).

Notes

1 *These tests apply to all vehicles fitted with obligatory headlamps.*

2 *With some types of vehicles fitted with hydropneumatic suspension systems it is necessary to have the engine idling while checking headlamp aim.*

3 *Vehicles having a four headlamp system must provide the dipped (passing) beam from the outer lamps. The aim of these lamps must be checked by the method appropriate to their type.*

4 *These methods of inspection relate to the use of beam checking equipment having a collecting lens.*

5 *For maximum illumination of the road and therefore for safety, it is important that headlamps should not be aimed too low or too far to the nearside.*

Method of inspection	Principal reasons for failure	Notes
of the headlamp under test in accordance with the equipment manufacturer's instructions.	(iii) The area of maximum intensity is wholly to the offside of the vertical line of the aiming screen, that is the lamp or bulb is for a right hand rule of the road vehicle.	
6.3 Following the instructions given by the manufacturer for the particular headlamp aim equipment being used, determine on the equipment screen:	3 European Type Headlamp Checked on dipped (passing) beam. (See page I/9). (i) The position of the horizontal boundary line between the high intensity and low intensity areas on the offside of the image is less than 0.5° below the horizontal line of the aiming screen, (Diagram 9).	
6.3.1 For headlamps checked on main (driving) beam, the position of the centre of the maximum light intensity area relative to the screen line markings. (See page I/7 Diagram 3).	(ii) The breakpoint or kink of the image (ie the intersection between the horizontal part of the cut off line and the 15° kick up on the nearside of the image) is to the offside of the vertical line of the aiming screen. (Diagram 9). (See Note 5 on maximum illumination).	
6.3.2 For headlamps checked on dipped (passing) beam, the position of the edges of the maximum light intensity area relative to the screen line markings. (See page I/8 Diagram 6), or	(iii) The rising part of the image (ie the 15° kick up) is to the offside of the vertical aiming screen line, that is the lamp fitted is for a right hand rule of the road vehicle.	
6.3.3 The position of the boundaries of the high intensity and low intensity areas and the position of the break point, relative to the screen line markings. (See page I/9 Diagram 9).		
6.4 If the above operations have been carried out with the headlamp emitting main (driving) beam, switch the headlamp to dipped beam and observe that the brightest part of the image moves downwards or downwards and to the nearside.		
6.5 If a headlamp designed to emit one type of beam does not illuminate when that beam is switched on, or in the case of a headlamp designed to emit both main and dipped beam which does not illuminate when both or either of the beams is switched on, proceed as detailed in subsection 2 (page 2).		

BRITISH AMERICAN TYPE HEADLAMP CHECKED ON MAIN (DRIVING) BEAM

The lens of this type headlamp is always circular and the lamp is likely to be of sealed beam construction. The lens may or may not be marked with the figure 1 and an arrow to indicate the direction of dipping.

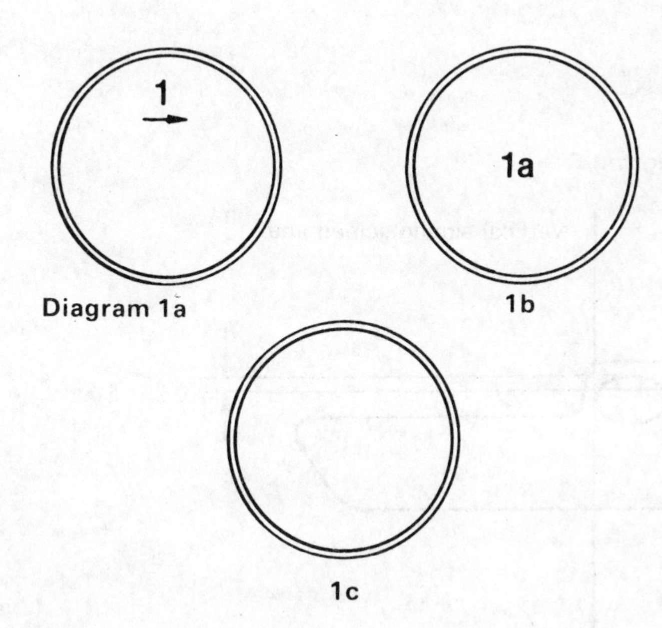

Diagram 1a 1b 1c

Note: For additional lens markings see page I/10

British American main beam image

Diagram 2

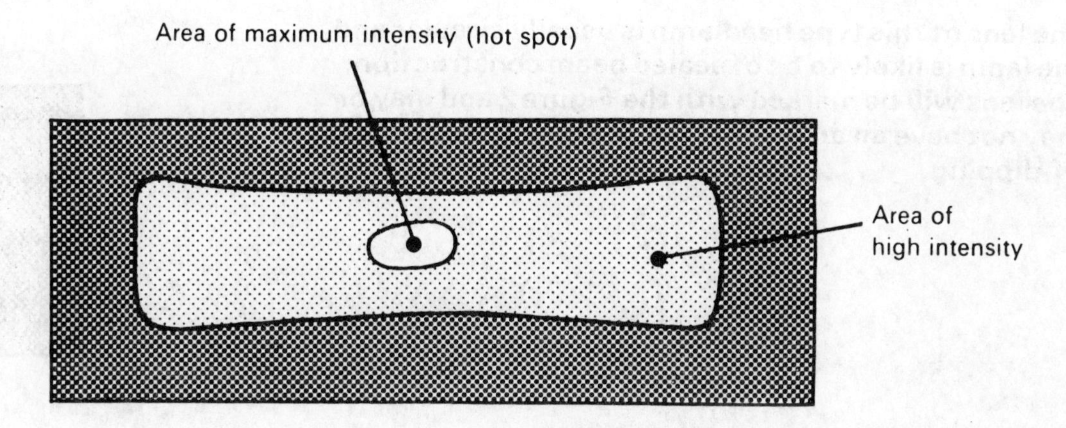

Diagram 3

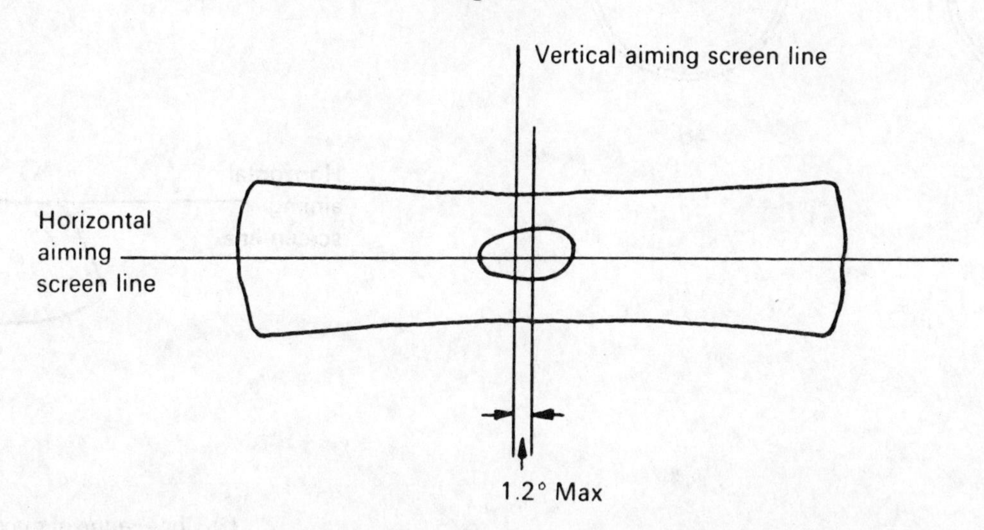

(i) The centre of the area of maximum intensity must not be above the horizontal line of the aiming screen.

(ii) The centre of the area of maximum intensity must not be more than 1.2 to the offside of the vertical line of the aiming screen.

BRITISH AMERICAN TYPE HEADLAMP CHECKED ON DIPPED (PASSING) BEAM

The lens of this type headlamp is usually circular and the lamp is likely to be of sealed beam construction. The lens will be marked with the Figure 2 and may or may not have an arrow marked to show the direction of dipping.

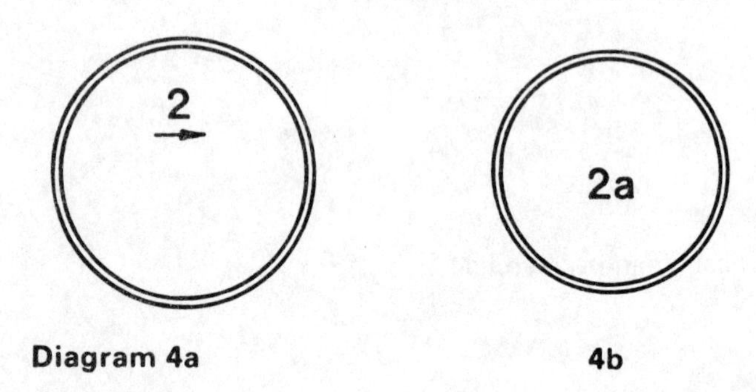

Diagram 4a **4b**

Note: For additional lens markings see page I/10

British American dipped beam image

Diagram 5

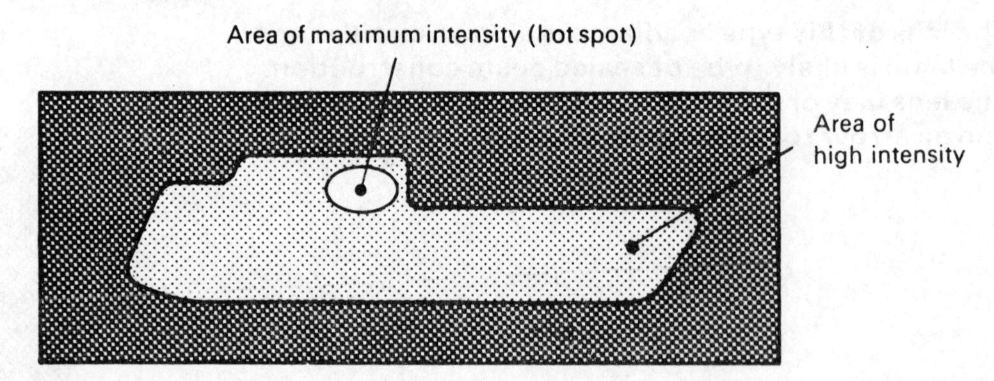

Diagram 6

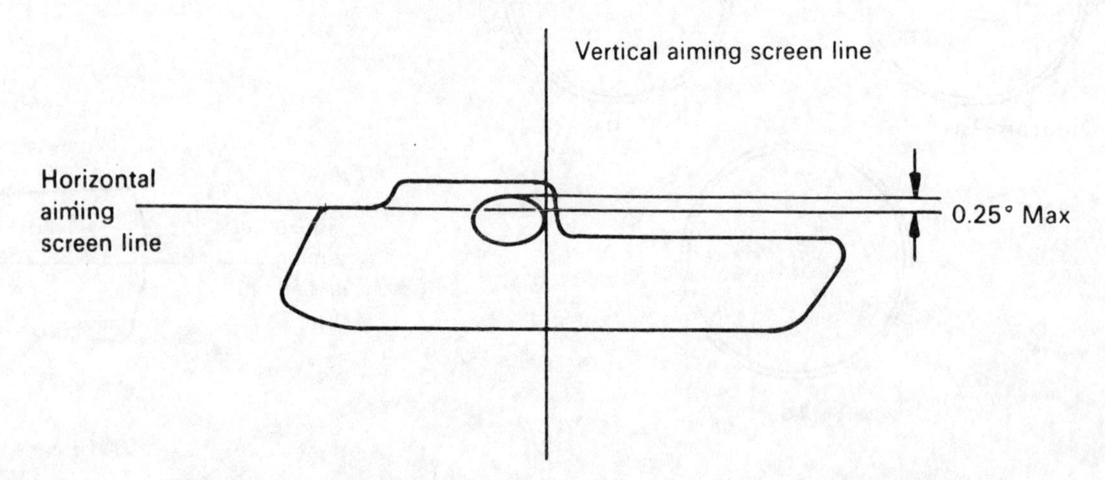

(i) Upper edge of maximum intensity area must not be more than 0.25° above the horizontal line of the aiming screen.

(ii) Offside edge of maximum intensity area must not be to the offside of the vertical line of the aiming screen.

EUROPEAN TYPE HEADLAMP CHECKED ON DIPPED (PASSING) BEAM

The lens of this type headlamp may be circular, rectangular or trapezoidal in shape and has a segment shaped pattern moulded into the glass. The lens is normally marked with a figure 2 and an arrow indicates the direction of dipping.

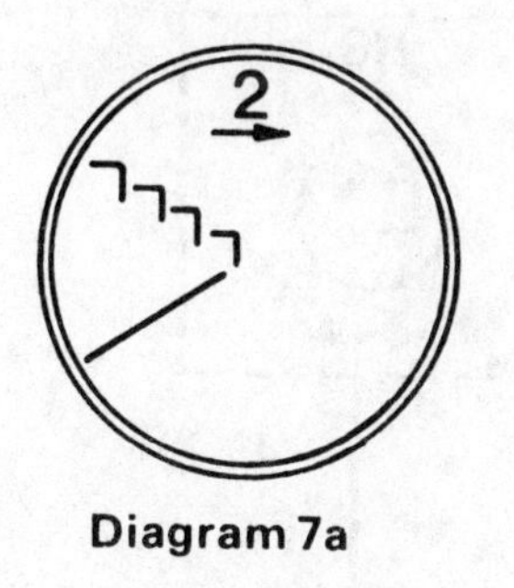

Diagram 7a

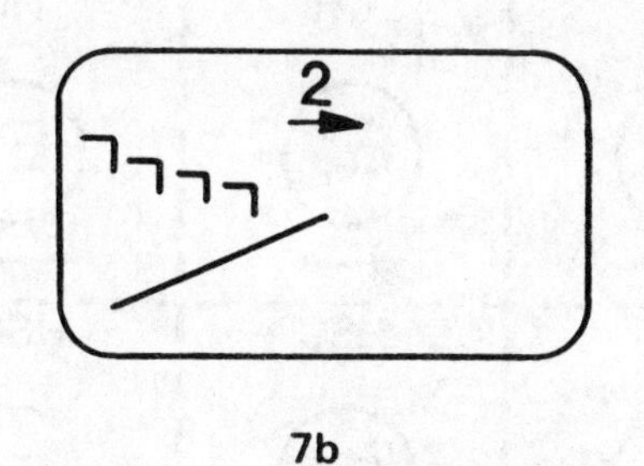

7b

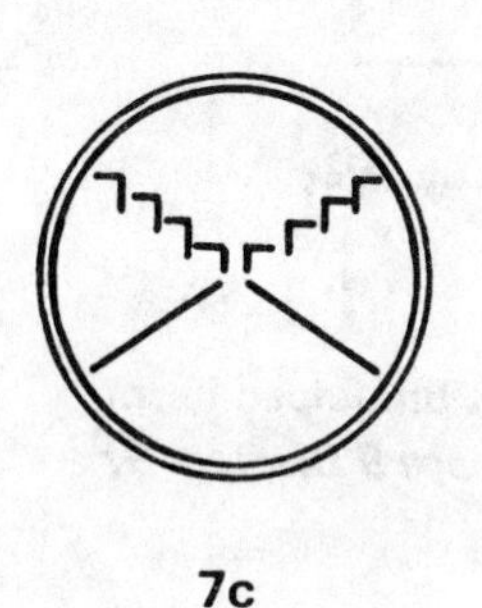

7c

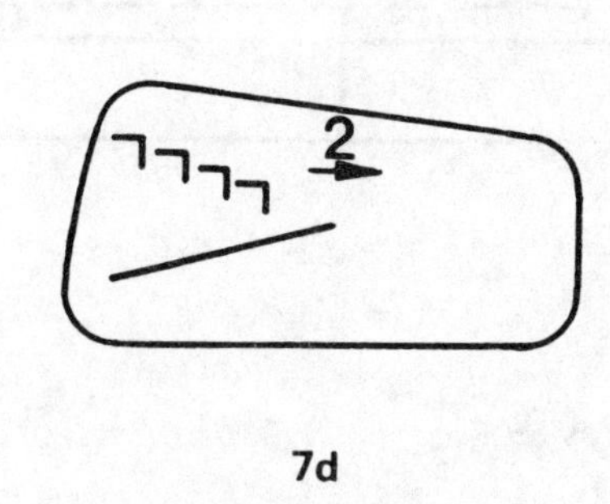

7d

Note: For additional lens markings see page I/10

European type dipped beam image

Diagram 8

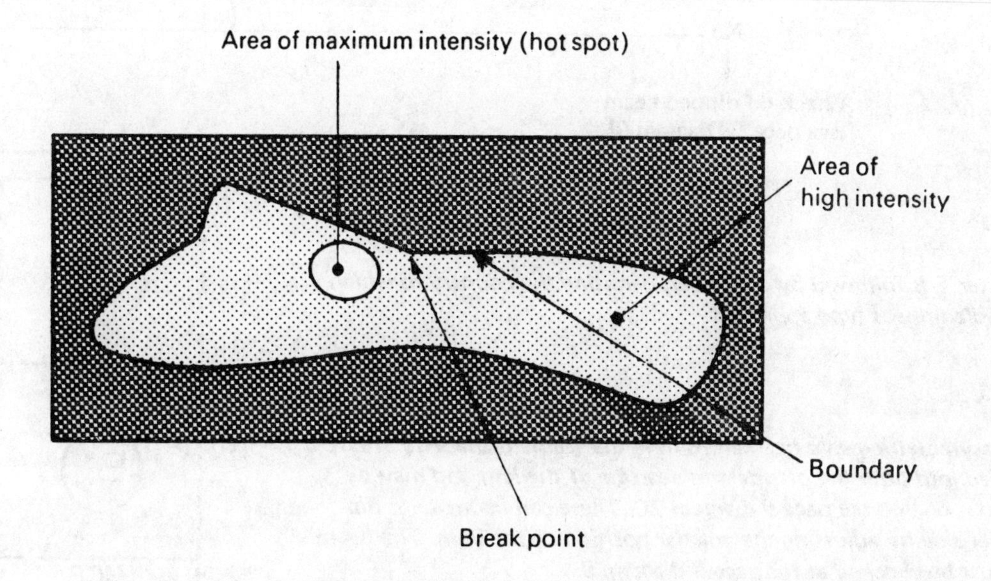

Diagram 9

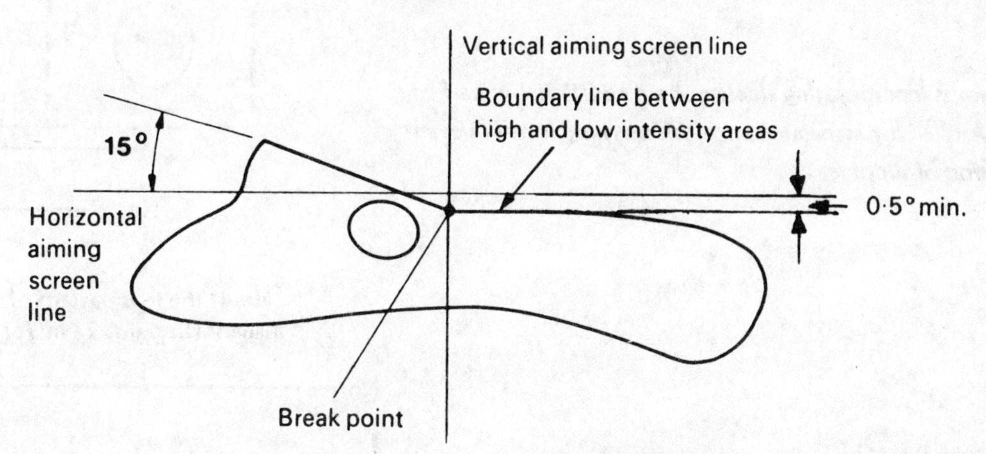

(i) Boundary line between high and low intensity areas to be not less than 0·5° below horizontal line of the aiming screen.

(ii) Break point not to be to the offside of the vertical line of the aiming screen.

HEADLAMP TYPES, LENS MARKINGS AND METHODS OF TEST

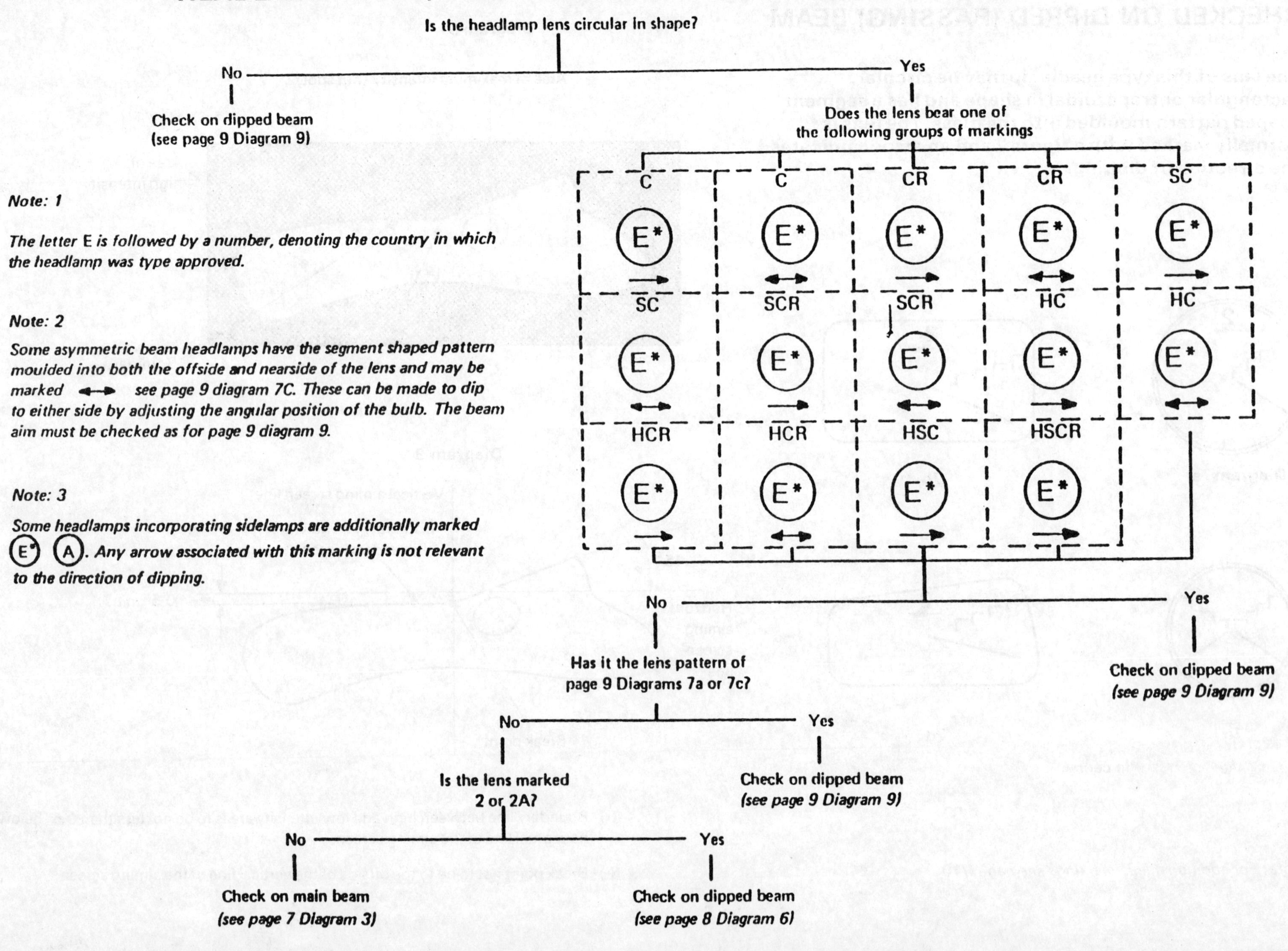

Note: 1

The letter E is followed by a number, denoting the country in which the headlamp was type approved.

Note: 2

Some asymmetric beam headlamps have the segment shaped pattern moulded into both the offside and nearside of the lens and may be marked ↔ see page 9 diagram 7C. These can be made to dip to either side by adjusting the angular position of the bulb. The beam aim must be checked as for page 9 diagram 9.

Note: 3

Some headlamps incorporating sidelamps are additionally marked (E) (A). Any arrow associated with this marking is not relevant to the direction of dipping.*

Section II

Steering (Including Suspension)

Method of inspection	Principal reasons for failure	Notes
1 STEERING WHEEL AND COLUMN		
1.1 Before carrying out this inspection, ensure that any mechanism for adjusting the steering column is fully locked. Also, only exert reasonable pressure on the steering wheel, particularly with a collapsible steering column. It is also essential to be satisfied that any excessive movement is due to wear or deterioration and not due to the design of the mechanism. Rock the steering wheel from side to side at right angles to the column, whilst applying slight downward and upward pressure. See or feel if there is any relative movement between the steering column shaft and steering wheel, indicating looseness between the two, or excessive wear in the column bush.	1 Any relative movement between the steering wheel and steering column indicating movement in the shaft splines or taper. 2 Any abnormal movement indicating excessive wear in the column top bearing.	1 *It is essential to be satisfied that excessive movement is due to wear or deterioration and not due to the design of the mechanism.* 2 *It may be necessary to open the bonnet, or to examine from inside the vehicle, some steering column universal joints and clamps. Similarly, in order to identify the cause of excessive movement, it may be necessary to examine the mechanism whilst another person carries out the functions detailed in the "Method of Inspection" column.*
1.2 Pull and push the steering wheel in line with the steering column. See if there is any movement at the centre of the steering wheel, indicating wear in the steering box or column bearing and/or a worn flexible coupling.	3 Appreciable movement of the centre of the steering wheel up and down which can be identified as being caused by an excessively worn steering box or column thrust bearing, or by a badly deteriorated flexible coupling, or a worn universal joint or upper column assembly bearings.	

	Method of inspection		Principal reasons for failure		Notes
1.3	Push the wheel away from and pull it towards the body. See if there is any movement of the column due to an insecure column mounting bracket.	4	Excessive movement of the top of the steering column from its longitudinal axis, due to an insecure or fractured top mounting bracket.		
1.4	Examine the steering wheel hub, spokes and rim for fractures or loose spokes.	5	A fractured spoke, hub or rim.	3	*Cracks or incompleteness of the covering skin of a steering wheel or hub, are not grounds for failure.*
		6	A spoke loose at the hub or rim.		

2 STEERING MECHANISM

	Method of inspection		Principal reasons for failure		Notes
	This sub section deals with the inspection of the steering mechanism for all types of front suspension. The inspection is carried out with the vehicle over a pit or on a raised hoist, with the steered wheels in the straight ahead position, supporting the weight of the vehicle.				
2.1	Whilst the steering wheel is rocked gently in each direction, to a point where movement of the drop arm or steering rod is just felt or seen, note the amount of movement at the circumference of the steering wheel.	1	Excessive free play at the steering wheel. The amount of acceptable free play will depend on the type of steering mechanism (ie steering box or rack and pinion) and the diameter of the steering wheel. As a general rule free play in excess of 3" should be considered excessive for vehicles with steering box mechanisms, whilst for rack and pinion mechanisms free play in excess of ½" should be considered as excessive.	1	*It may be necessary to open the engine compartment to examine some steering components on some vehicles.*
2.2	Whilst the steering wheel is rocked firmly in each direction against the resistance of the tyres, with a force sufficient to load the steering mechanism and joints, carry out a visual examination of the complete steering mechanism (See Note 2). The various points to look for are as detailed below:		Any of the conditions listed below which, in the opinion of the tester, are such that the steering of the vehicle may be impaired to such an extent that directional control of the vehicle may be unpredictable. (See Notes).	2	*It is essential that for the inspection detailed in subsection 2.2 the steering is moved to load the steering joints, and not merely to take up any free play as for the inspection detailed in subsection 2.1.*
				3	*This inspection must be carried out without any dismantling. It is therefore accepted that it is not always possible to examine completely items which are covered with protective gaiters.*
2.2.1	Relative movement between the steering box or steering rack assembly and the chassis frame or body shell.	1	Excessive play in a steering joint.		
		2	Excessive stiffness in the steering.		
2.2.2	Insecurity between the various parts of the steering box or steering rack assembly.	3	Deterioration of the bushing material in a bushed joint, which results in excessive play.		

Method of inspection

2.2.3 Relative movement between any idler arm attachment and the chassis frame or body shell.

2.2.4 Relative movement between the drop arm and the sector shaft.

2.2.5 Wear or end float in a sector shaft or bearing. Wear or end float in an idler arm or bearing.

2.2.6 Wear in all ball or other type joints in the steering mechanism. Care must be taken to distinguish between relative movement due to excessive wear and relative movement due to built-in clearance or spring loading of a joint. If excessive wear is suspected in a joint, apply pressure to the joint carefully by means of a pinch bar to assess more accurately the reason for the relative movement.

2.2.7 The security of all ball pin shanks and nuts.

2.2.8 Relative movement between steering arms and stub axles, indicating that they are not properly secured.

2.2.9 The security of all nuts and retaining devices, including track rod ends and drag link ends. The alignment of ball pins to see that they are correctly located relative to their housings.

2.2.10 Fractures or severe deformation of the mechanism.

2.2.11 Fluid leakage from any steering damper seal or gland.

2.2.12 External damage or serious corrosion to the damper body or cover.

Principal reasons for failure

4 Relative movement between a sector shaft and drop arm.

5 Excessive wear in a pivot point.

6 A track rod end or drag link end loose, or misaligned with its ball.

7 A ball pin shank loose.

8 Sharp or deep grooves at the neck of a ball pin.

9 Insecurity of any part fixed to the vehicle structure (eg steering box, rack housing or pivot bearing housing).

10 A part of the steering box or rack and pinion assembly insecure.

11 Relative movement between a steering arm and stub axle.

12 A steering component fractured or deformed.

13 A steering component which is obviously unsuitable for the function it is required to perform (eg a repair or replacement carried out by the use of a component which is not suitable for the steering mechanism).

14 A retaining device missing or insecure.

15 A locking device missing or insecure.

16 A sector shaft twisted.

Notes

4 *Bonded joints will show movement due to elasticity or slight deterioration. This is acceptable.*

5 *Wear or play must only be regarded as excessive if it is clear that the component is at the stage when replacement, repair or adjustment is necessary.*

Method of inspection	Principal reasons for failure	Notes
	17 Excessive play or end float of a sector shaft or idler arm shaft.	
	18 A steering box or rack and pinion housing bolt which is loose or missing.	
	19 A steering box or rack and pinion housing fractured.	
2.2.13 The security of attachment of the damper to the chassis frame or body shell and to the steering linkage.	20 Fluid leakage from a steering damper gland to such an extent that it is clear that the gland has failed.	
	21 External damage or serious corrosion to the damper body or cover such that it is clear that the movement of the steering linkage could be impeded in a manner not intended by the design.	
	22 A steering damper missing.	
	23 A steering damper insecurely fixed to the chassis frame, body shell or steering linkage.	
2.2.14 The condition of the chassis frame and body shell for excessive corrosion or fractures in the vicinity of the steering box or rack and pinion housing, the idler arm mounting, the steering damper fixing points and wishbone anchorages.	24 Excessive corrosion, severe distortion or fractures in a load bearing member of the vehicle structure or surrounding panelling within 30 cm (12") of the steering box, rack and pinion housing, any idler arm mounting, steering damper fixing points or wishbone anchorage.	
2.2.15 For front wheel drive vehicles, the condition of the drive shaft inner and outer universal joint couplings for security and damage. (See Note 1).	25 A drive shaft constant velocity or universal joint coupling excessively worn or insecure.	6 *In order to check the condition of drive shaft universal joints it may be necessary to chock the vehicle, allowing the vehicle some movement, release the handbrake and engage reverse gear and rock the vehicle backwards and forwards.* *IMPORTANT: The ignition must be switched off during this operation.*
	26 A flexible rubber or fabric universal coupling unit damaged by severe cracking or breaking up.	
	27 A flexible rubber or fabric universal coupling unit softened by oil contamination, or insecure.	
	28 An insecure or fractured U bolt securing a joint bearing.	

Method of inspection		Principal reasons for failure		Notes
3 POWER STEERING				
3.1	With the vehicle in neutral gear, the parking brake applied and the engine running, rock the steering wheel gently in both directions whilst the following are examined visually:	1	The power steering does not operate correctly.	
		2	A locking device insecure or missing.	
		3	Excessive play in a power steering mechanism joint.	
3.1.1	All parts of the steering linkage associated with the power steering mechanism, as detailed in subsection 2.2 above;	4	Excessive deterioration in any bushing material of a joint, resulting in excessive relative movement between the joint components.	
3.1.2	Any hydraulic fluid hoses and unions for leaks, taking care to distinguish between leaks from the power steering and extraneous fluid which may not necessarily be a leak from that mechanism;	5	Insecurity of any part fixed to the vehicle structure.	
		6	A damaged hose or any fluid leak.	
3.1.3	The condition of the power steering pump drive and the security of the pump mounting.	7	Insecure drive or drive belt to the pump.	
		8	The pump mounting insecure.	
4 FRONT WHEEL BEARINGS				
4.1	With the vehicle over a pit or on a raised hoist, jack up the front wheels (see Figs. page 24 for details of jacking the various types of suspensions).			
4.2	While the front wheels are being jacked up, look for movement in the inner wishbone bearings, as the direction of loading changes.	1	Excessive wear in a pin or wishbone bearing or serious deterioration of a flexible rubber bush.	
4.3	When jacking is completed proceed as follows:			
4.3.1	Spin each front roadwheel in turn and listen for any sound indicating roughness in the wheel bearing.	2	Roughness or tightness in a wheel bearing whilst the wheel is rotating, indicating imminent failure of the bearing.	

Method of inspection		Principal reasons for failure		Notes	
4.3.2	With the wheel stationary, feel by moving the wheel gently for any tightness or excessive play, indicating incorrect adjustment of the bearing. This should be done with the wheel in at least two positions.	3	Excessive play or insufficient clearance in a wheel bearing.		
4.4	For front wheel drive vehicles, carry out the following inspections:				
4.4.1	With the vehicle in neutral gear and while rotating the wheels when they are on both locks, examine visually the gaiters of the constant velocity joints while opening out the pleats with the fingers.	4	A split or missing gaiter or a gaiter which is insecurely mounted to its housing.		
4.4.2	Check the front wheel drive shafts for straightness and damage.	5	A damaged or bent shaft.		

5 SUSPENSION (Figs. page 24)

Method of inspection		Principal reasons for failure		Notes	
5.1	For suspensions of the types shown in figures 1 and 2, jack up the front suspension so that the roadwheels are clear of the ground and the suspension is as near as possible to normal running height.	1	A kingpin loose in its mounting boss or stub axle bush, indicating excessive wear.	1	*It may be necessary to open the engine compartment to examine some suspension components on some vehicles.*
		2	A kingpin retaining device loose or missing.		
	Whilst each wheel is held at the top and bottom and rocked, examine for movement between the kingpin and its bushes or its mounting in the axle boss, movement between the wishbone outer suspension ball joints and their housings and movement in the upper inner wishbone bearings.	3	Excessive wear in any suspension swivel pin or suspension ball joint.		
		4	Excessive wear in a pin or wishbone bearing.		
		5	Serious deterioration of a flexible bush.		
5.2	With a bar under each front roadwheel attempt to lift the wheel, observe any movement between the stub axle yoke and its housing at the thrust bearing, or between suspension ball joints and their housings.	6	Excessive lift between a stub axle and axle housing, such that damage to, or early failure of, the thrust washer or bearing is likely to occur.	2	*Some lower suspension ball joints designed as thrust bearings are not internally spring loaded and have a considerable amount of built-in clearance. This clearance becomes apparent only when the suspension load is relieved by jacking.*
		7	Excessive play caused by wear in a ball joint. (See Note 1).		

Method of inspection		Principal reasons for failure		Notes	
5.3	Examine beam axles, wishbones and stub axles for damage and distortion.	8	A distorted axle beam or component.		
		9	A distorted, excessively corroded or fractured wishbone arm.		
5.4	Examine the condition of the chassis frame and body shell structure in the vicinity of suspension mounting points and suspension subframe mounting points for fractures, excessive corrosion and distortion.	10	Excessive corrosion, deformation or fracture in a load bearing member of the vehicle structure and surrounding panelling within 30 cm (12") of a suspension or subframe mounting point.		

6 ALL SUSPENSION TYPES (Figs. page 24)

Method of inspection		Principal reasons for failure		Notes	
6.1	Lower the front wheels so that they bear the weight of the vehicle and are resting on greased plates, turntables or other means which enable the wheels to be turned from lock to lock without strain.				
6.2	With the front wheels turned from lock to lock, as required, carry out the following examination:	1	A component of the steering mechanism, roadwheels or tyres fouling any fixed part of the vehicle structure, including flexible brake hoses.		
6.2.1	Determine whether there is any fouling, particularly whether the flexible brake hoses are fouling any fixed or moving part.	2	An incorrectly adjusted lock stop.		
		3	A lock stop loose, damaged or insecurely locked.		
6.2.2	Examine steering lockstops for security and correct adjustment.	4	A steering rack gaiter insecure, split or missing.	1	Steering rack gaiters should be spread with the fingers for proper examination.
6.2.3	Examine steering rack gaiters for security and condition.	5	Excessive tightness or roughness in the steering mechanism.		
6.2.4	Assess whether there is any tightness or roughness in the steering mechanism.	6	A metal or flexible brake hose stretched, twisted or seriously damaged by fouling the wheel, tyre or part of the suspension.		
6.2.5	Examine the condition of wishbones and their inner bearings, track control arms, suspension radius rods and their mounting bushes or washers.	7	Excessive wear in a pin or wishbone bearing or serious deterioration of a flexible bush.		

Method of inspection		Principal reasons for failure	Notes
6.2.6	Examine the condition of each suspension spring, displacer or bellows for correct mounting and for damage.	8 A distorted, excessively corroded or fractured wishbone.	
6.2.7	Examine the condition of each front shock absorber as detailed in subsection 10.	9 A distorted, excessively corroded or fractured track control arm.	
		10 A distorted, insecure or fractured radius rod.	
		11 A seriously deteriorated radius arm, flexible bearing or washer.	
		12 A radius arm nut insecurely locked.	

7 SUSPENSION TYPE FIG 3 (Page 24)

Method of inspection		Principal reasons for failure	Notes
7.1	Shake each roadwheel vigorously to determine the condition of the outer suspension ball joint assembly for play between the ball and the housing.	1 A seriously worn suspension ball joint assembly.	
		2 A ball joint securing nut not tight.	
		3 A ball joint nut not locked.	
		4 A seriously worn pin or bush in an inner wishbone bearing.	

8 SUSPENSION TYPE FIG 4—MACPHERSON STRUT TYPE (Page 24)

Method of inspection		Principal reasons for failure	Notes
8.1	While each front wheel is shaken vigorously, examine each suspension strut for wear at the strut sliding bush and gland as well as for movement at the strut upper support bearing. Check for leakage of fluid from the gland, serious damage or excessive corrosion of the strut casing, wear in the rod and the condition of the bonding between the metal and flexible material in the strut upper support bearing.	1 Excessive wear on a shock absorber strut and/or bush.	
		2 Excessive leakage of fluid from a shock absorber gland.	
		3 Excessive play in a strut upper support bearing assembly.	
		4 Roughness or stiffness in a strut upper support bearing.	
		5 Serious external damage or excessive corrosion of a strut casing.	

Method of inspection		Principal reasons for failure		Notes	
8.2	Whilst each front wheel is shaken vigorously (grasping it at the "3 o'clock" and "9 o'clock" positions) determine the condition of the outer ball joints and track control arm inner bushes, for movement indicating the degree of wear.	6	Serious deterioration of the bonding between metal and flexible material of an upper support bearing.		
		7	A loose or insecurely locked nut in the upper support bearing assembly.		
		8	Serious deterioration of the bonding or flexible material of a track control arm or radius member bush or mounting.		
		9	Excessive wear in a strut lower ball joint.		
		10	A ball joint or assembly cover nut loose, or insecurely locked.		
8.3	Examine the condition of each coil road spring for correct mounting and damage. (See subsection 9.2).				
8.4	Examine the vehicle structure in the vicinity of each upper strut support bearing mounting, track control arm or radius arm mounting, for fractures, corrosion and distortion.	11	Excessive corrosion, deformation or fracture in a load bearing member of the vehicle structure within 30 cm (12") of an upper strut support bearing mounting, a track control arm mounting or radius arm mounting.		

9 SUSPENSION ASSEMBLIES—SPRINGS, TORSION BARS, ETC

(This subsection applies to front and rear suspensions and the inspections are carried out with the vehicle on a pit or raised hoist).

9.1 Sub-frames

9.1.1	Examine the condition of each sub-frame for fractures, excessive corrosion and distortion.	1	A sub-frame badly distorted in such a manner as to affect its strength or operation.	*1*	*This inspection is to be carried out without any dismantling. It is, therefore, accepted that it is not always practicable to inspect these items completely. However, the inspection should be carried out on those parts which are accessible and which can be seen from underneath the vehicle.*
9.1.2	Examine the mounting of each sub-frame to the chassis frame and to the wheel assembly, looking particularly for any insecure mounting, any fractures or any excessive corrosion to	2	A sub-frame fractured.		
		3	A sub-frame which is so corroded that it is clear that its strength has been materially reduced.		

Method of inspection	Principal reasons for failure	Notes
the vehicle structure in the vicinity of the sub-frame mountings. Examine bonded flexible mountings for deterioration.	4 A fractured or corroded sub-frame inadequately repaired.	
	5 An insecurely locked or defective mounting.	
	6 A fracture or excessive corrosion of a load bearing member of the vehicle structure within 30 cm (12") of any sub-frame mounting.	
	7 A badly deteriorated flexible mounting.	
9.2 Coil Spring or Displacer Units		
9.2.1 Examine the condition of each coil spring or displacer unit for correct mounting and damage. Examine the vehicle structure in the vicinity of spring mountings for fractures, excessive corrosion and distortion. Examine any interconnecting pipes between displacer units.	1 A coil spring incomplete or fractured, or with a cross section so reduced by wear or corrosion that its strength is seriously reduced.	
	2 A coil spring or displacer unit which is not properly seated.	
	3 Inadequate clearance of the axle or suspension with the bump stop or chassis.	
	4 Insecure, damaged, leaking or corroded interconnecting pipes between displacer units.	
	5 An excessively corroded, fractured or distorted load bearing member of the vehicle structure or panelling within 30 cm (12") of any spring mounting.	
9.3 Leaf Springs		
9.3.1 Examine each spring for fractures, displaced leaves and for camber. See that each spring is fitted so that the axle is correctly located and that it is secure to the axle.	1 A fractured leaf in a spring.	1 *This inspection is to be carried out without any dismantling. It is therefore accepted that it is not always practicable to inspect these items completely. However, the inspection should be carried out on those parts which are accessible and which can be seen from underneath the vehicle.*
	2 Splaying of leaves of a spring to such an extent that other parts of the vehicle are fouled, or the efficiency of the spring is seriously impaired.	
9.3.2 Examine each spring anchor bracket and spring shackle bracket and the associated pins and bushes for wear, security and adequate	3 Inadequate clearance of the axle or suspension with the bump stop or chassis.	

Method of inspection	Principal reasons for failure	Notes
locking. Pay particular attention to the amount of side play in the brackets. A small pinch bar will be required for this inspection to check the wear, but where this cannot be used a visual examination should be made.	4 A broken centre bolt.	
	5 A spring fitted in such a way that the axle is incorrectly located.	
	6 Looseness of a bolt or plate securing a spring to an axle.	
9.3.3 Examine the condition of each shock absorber as detailed in subsection 10.	7 A defective spring-eye.	
9.3.4 Examine the chassis or body structure in the vicinity of each spring mounting for fractures, excessive corrosion and distortion.	8 An anchor pin and/or bush excessively worn.	
	9 A shackle pin and/or bush excessively worn.	
	10 An anchor pin loose in its bracket.	
	11 A shackle pin loose in its bracket.	
	12 A shackle or anchor pin locking device which is insecure or missing.	
	13 Excessive corrosion, serious distortion or fracture in a load bearing member or surrounding panelling of the chassis or body structure within 30 cm (12") of any spring mounting.	
9.4 Suspension Locating Links, Radius Arms, Tie Bars, etc		
9.4.1 Examine each radius arm/link for distortion, fracture or excessive corrosion. See if there is any wear in the mountings and whether the arm mounting or pin is securely fixed to the vehicle structure.	1 A distorted, fractured or excessively corroded radius arm or locating link.	*1 This inspection is to be carried out without any dismantling. Therefore, it is not always practicable to inspect these items completely. However, the inspection should be carried out on those parts which are accessible and which can be seen from underneath the vehicle.*
	2 Excessive wear in a radius arm or locating link bush or bearing.	
	3 A radius arm or locating link mounting or pin, which is insecure or not properly locked.	
	4 Serious deterioration of the bonding or flexible material of a radius arm or link.	

Method of inspection	Principal reasons for failure	Notes
9.4.2 On vehicles which have a drive shaft which forms part of the suspension, examine the shaft for distortion, damage and serious corrosion, the universal joint bearings for wear and the flanges and bolts for security.	5 A distorted, damaged or excessively corroded drive shaft. 6 An excessively worn universal joint bearing. 7 An incorrectly seated universal joint flange. 8 A flange bolt loose or inadequately locked.	
9.4.3 Examine the condition of each shock absorber as detailed in subsection 10.		
9.4.4 Examine the vehicle structure and panelling in the vicinity of arm/link mountings for fractures, excessive corrosion and distortion.	9 Excessive corrosion, serious distortion or a fracture in a load bearing member of the vehicle structure or panelling within 30 cm (12") of a radius arm or locating link mounting.	
9.5 Torsion Bar Suspension		
9.5.1 Examine each torsion bar for fractures, excessive corrosion and pitting. Using a pinch bar as necessary, see whether there is any excessive free play in the attachment of the torsion bar to the suspension arm or wishbone or guide. Examine the security of the torsion bar abutment screw assembly to the body structure of the vehicle.	1 A fracture or excessive corrosion in a torsion bar which is likely to cause failure of the bar. 2 Excessive free play in the attachment of the torsion bar to the suspension arm or wishbone. 3 A damaged or inadequately locked torsion bar abutment. 4 Inadequate clearance of the axle or suspension with the bump stop or chassis. 5 Excessive corrosion, serious distortion or fracture in a load bearing member of the vehicle structure or panelling within 30 cm (12") of a torsion bar attachment.	
9.5.2 Examine the condition of each shock absorber as detailed in subsection 10.		
9.6 Anti-roll Bars		
9.6.1 Examine each anti-roll bar for distortion, fractures or serious corrosion. See if there is any wear in the mountings and whether the	1 A badly distorted, fractured or excessively corroded anti-roll bar.	

Method of inspection		Principal reasons for failure		Notes	
	bars are securely fixed. Examine the vehicle structure and panelling in the vicinity of the mountings for fractures, excessive corrosion and distortion.	2	Serious deterioration of the bonding or flexible material of an anti-roll bar joint or bearing.		
		3	Excessive corrosion, serious distortion or fracture in any load bearing member of the vehicle structure or panelling within 30 cm (12") of any anti-roll bar mounting.		

10 SHOCK ABSORBERS

Method of inspection		Principal reasons for failure		Notes	
	(This subsection applies to front and rear shock absorbers and the inspection is carried out with the vehicle on a pit or over a raised hoist.)				
10.1	Examine each shock absorber for damage, corrosion, fluid leaks and security of attachment. For shock absorbers incorporated in strut type suspensions carry out the inspection as detailed in subsection 8.	1	Fluid leakage to such an extent that it is clear that the fluid seal of a shock absorber has failed.	*1*	*With fluid leaks it is necessary to ensure that any fluid in the vicinity of the unit is from the shock absorber and not from any extraneous source.*
		2	External damage or corrosion to the casing of a shock absorber to such an extent that the unit does not function.	*2*	*The inspection does not include the removal of dust sheilds, if these are fitted.*
10.2	Check whether a shock absorber is missing.	3	A shock absorber insecurely fixed.	*3*	*Where flexible dust shields are fitted, the existence of fluid leaks from a shock absorber can frequently be detected by squeezing the dust shield and watching for fluid leaking from under it.*
10.3	Carefully pull or push down the vehicle at each corner and release it. Note the rebound of the body to determine whether each shock absorber is producing a damping effect.	4	A shock absorber lever or link insecurely attached.		
		5	A shock absorber missing.	*4*	*Some vehicles have shock absorbers as optional fittings. If fitted these must meet the requirements.*
		6	A shock absorber which has no damping effect on the suspension.		

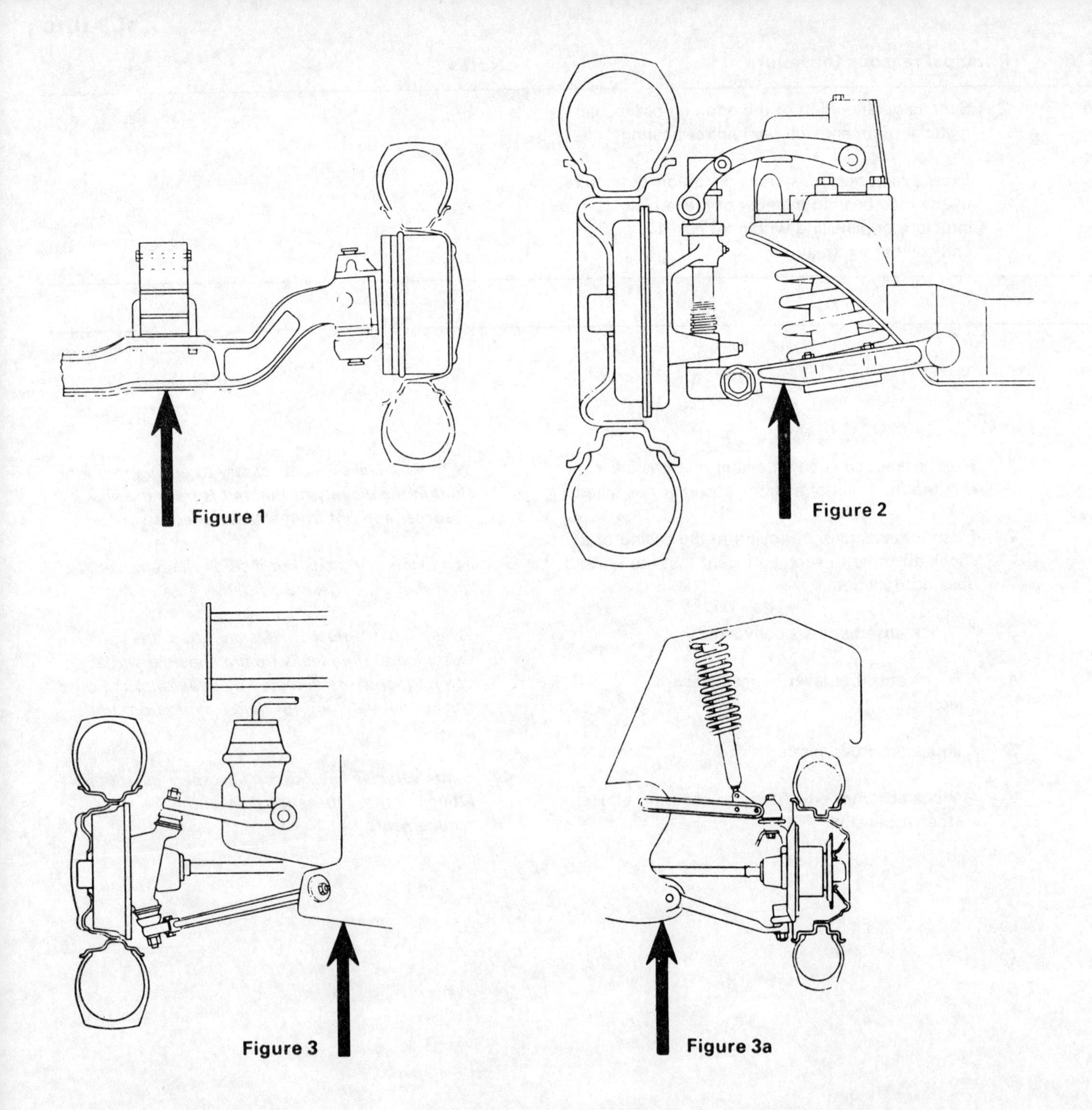
Figure 1
Figure 2
Figure 3
Figure 3a

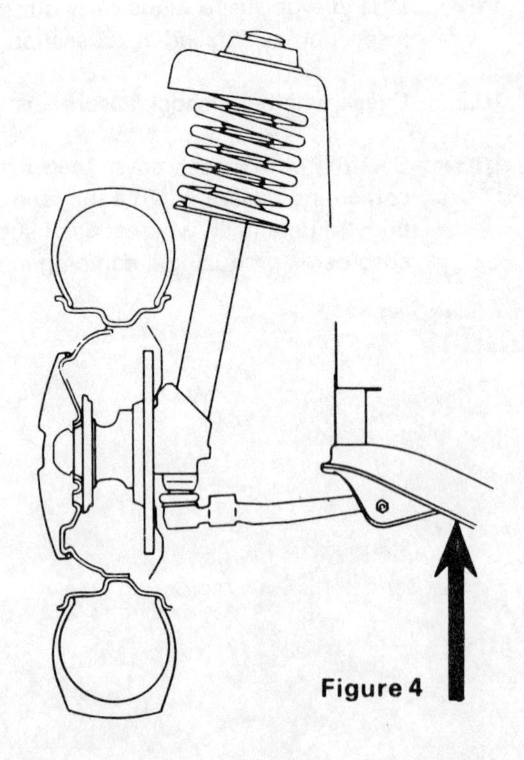
Figure 4

Section III

Brakes

Method of Inspection		Principal reasons for failure		Notes	
1	**PARKING BRAKE OPERATING LEVER**				
1.1	Whilst the parking brake lever is operated, check that the parking brake is designed to prevent at least two of the vehicle's roadwheels (or in the case of a three wheeled vehicle at least one of the roadwheels) from turning.	1	The vehicle does not have a parking brake designed to prevent at least two of the roadwheels (or in the case of a three wheeled vehicle at least one of the roadwheels) from turning.	*1*	*This inspection applies to all vehicles.*
				2	*In this section it is assumed that the parking brake is applied by a hand operated lever (handbrake) and that the service brake is applied by a foot operated pedal (footbrake). On vehicles with foot operated and hand released parking brakes the "Method of Inspection" detailed will need to be varied for the particular mechanism concerned.*
1.1.1	Apply the parking brake and check that it can be set in the "applied" position. Determine the effectiveness of the holding-on mechanism by tapping the brake lever sideways and seeing that the lever does not release. With the holding-on mechanism released, move the lever sideways to check for wear in the pivot bearing.	2	With reasonable force applied to the lever, it cannot be set in the "hold-on" position.		
		3	The holding-on mechanism operates, but this is so worn, or the sideways movement of the lever is such, that the parking brake is likely to release inadvertently with small sideways force or accidental contact with the "hold-on" mechanism.	*3*	*An indication of the condition of the holding-on mechanism can be obtained by applying the parking brake and listening to the pawl mechanism operate on the ratchet.*
1.1.2	Where the mountings of the parking brake mechanism to the vehicle structure can readily be seen from inside the vehicle, inspect the security of the fixing and the condition of the body structure and panelling in this region.	4	The parking brake mechanism is not securely fixed to the vehicle structure.	*4*	*On some vehicles there is some sideways movement of the parking brake lever when new. Only movement which is such that the pawl is moved clear of the ratchet and the brake does not hold-on is to be regarded as a defect.*
		5	An excessively corroded, fractured or distorted load bearing member of the vehicle structure or panelling within 30 cm (12") of the handbrake lever mounting.	*5*	*If the security of the parking brake mechanism cannot be determined from inside the vehicle, this should be checked if possible, at the undervehicle inspection.*
		6	The travel of the parking brake lever is more than normal for the type of vehicle, before full		

Method of inspection		Principal reasons for failure		Notes
		resistance of the braking mechanism is felt, indicating that the mechanism requires adjustment. This will require confirmation when the parking brake performance test is carried out.	6	*If a parking brake requires adjustment, this will normally be indicated at the brake performance test.*
			7	*On some vehicles a complete inspection of the parking brake mechanism requires the bonnet to be opened.*

2 PARKING BRAKE MECHANISM (UNDER-VEHICLE)

Method of inspection		Principal reasons for failure		Notes
2.1	Whilst the parking brake lever is operated, examine as far as practicable the condition of the holding-on mechanism, the pivot point of the lever and the condition of the vehicle structure for corrosion in the vicinity of the brake lever attachment.	1	A parking brake holding-on mechanism which is worn to such an extent that the parking brake is likely to release inadvertently.	
		2	A parking brake lever pivot bearing which is worn to such an extent that the parking brake is likely to release inadvertently.	
		3	A parking brake lever attachment which is not securely fixed to the vehicle structure.	
		4	Excessive corrosion, fractures or deformation in any load bearing member or surrounding panelling of the vehicle structure within 30 cm (12") of a parking brake lever attachment pivot.	
2.2	Whilst the parking brake lever is operated from the fully "on" to the fully "off" position, examine the condition of all cables, rods, linkages, outer casings, guides, pivots, clevis pins, yokes and locking devices of the parking brake mechanism. Pay particular attention to those parts which are not readily visible when the parking brake is either in the fully "off" or in the fully "on" positions.	5	An operating cable which has been knotted. A recognised cable adjuster is acceptable, provided it does not foul other parts of the vehicle structure when the brake is operated.	
		6	A brake cable which is excessively corroded or frayed to such an extent that it is materially weakened.	
		7	Incorrect adjustment of the mechanism so that any lever or crank has inadequate reserve travel, or is pulled over centre to the extent that the application or performance of the brake is is restricted.	

Method of inspection		Principal reasons for failure		Notes
		8	Excessive corrosion, wear, chafing or distortion of any brake rod or assembly to an extent that its strength is materially weakened.	
		9	The absence or insecurity of any locking device.	
		10	An outer-cable stop which is worn or deteriorated to such an extent that the outer-cable passes through the stop, or that the stop itself is insecure.	
		11	Deteriorated cable or rod guides, which affect the operation of the brake.	
		12	Excessive wear in clevis pins, mounting pins, levers or pulleys in a compensating mechanism.	
3 SERVICE BRAKE OPERATING PEDAL				
3.1	From the driving seat, look for obvious signs of deterioration or incompleteness of the foot-brake pedal. Depress the pedal, at first slowly and then rapidly, to a point where sustained pressure can be held. Check whether the pedal creeps down from this point.	1	The service brake pedal is in such a condition that the brake cannot readily be applied.	
		2	When the service brake pedal is fully depressed, there is inadequate reserve travel for the pedal.	
		3	The pedal creeps down when sustained pressure is applied, indicating a leakage in the system.	
3.2	Check for obvious leaks of brake fluid in the vicinity of the master cylinder, particularly at the rear.	4	The pedal movement is spongy, indicating air in the hydraulic system.	
		5	There is an obvious leak at the master cylinder.	
4 SERVICE BRAKE MECHANISM UNDER VEHICLE				
4.1	With the vehicle over a pit or on a raised hoist, proceed as follows:	1	For all vehicles except veteran motor cars and three wheeled vehicles, two separate means of operating the brakes or two braking systems each with a separate means of operation, are not fitted.	*A veteran motor car is defined as one registered before 1 January 1915.*
4.1.1	For all vehicles, other than veteran motor cars and three wheeled vehicles, whilst the service brake is operated check there are two separate			

Method of inspection

means of operating the brakes, or two braking systems each with a separate means of operation are fitted. For veteran motor cars and three wheeled vehicles, check that the vehicle is equipped with at least one braking system.

4.1.2 With the service brake applied in order to pressurise the hydraulic system, examine all metal pipes, cylinders, reservoirs, flexible hoses and connections for leaks. Examine flexible hoses for bulges indicating weakness in their structure. Care must be taken during this inspection to distinguish between leaks from the braking system and extraneous fluid from another source.

4.1.3 Check the security of the master cylinder to the bulkhead and the condition of the servo.

4.1.4 Examine the condition and security of all visible metal pipes. Where corrosion or chafing has occurred, determine whether the resultant pitting or flaking has weakened the walls of the pipe or whether it is superficial.

4.1.5 Examine the condition of flexible brake hoses to see whether they are chafed, perished or peeling. Examine the outer sheath for rotting or fraying.

4.1.6 Examine, where practicable and without dismantling, the wear on brake pads or linings and the condition of brake discs and brake drums for fractures.

4.1.7 Examine, where practicable and without dismantling, brake discs or drums for contamination caused by leaking brake fluid, lubricating oil or grease.

Principal reasons for failure

2 For veteran motor cars and three wheeled vehicles, at least one braking system is not fitted.

3 A leak from a hose, pipe, connection, master cylinder or wheel cylinder unit.

4 A flexible hose which bulges under pressure.

5 A vacuum servo not functioning. (See Note 4).

6 The master cylinder mounting is insecure.

7 A metal pipe which is kinked or so insecure or chafed that failure is likely.

8 Corrosion or chafing of a metal pipe or unit which has resulted in deep grooving, pitting or flaking to such an extent that failure is likely.

9 A flexible hose which is twisted or has been stretched or is in such a condition that the inner wall is exposed, or the outer covering is chafed, cracked or peeling.

10 A brake disc or drum fractured or excessively scored, pitted or worn.

11 A brake pad or lining which is worn to the extent that replacement is necessary.

12 A brake disc or drum which is contaminated by brake fluid, lubricating oil or grease.

Notes

1 It is important to check the linkages and condition of any pressure controlling valve to the rear brakes when fitted.

2 For vehicles fitted with a cable operated service brake system, apply the same principles of inspection as are detailed in subsection 2 for the parking brake mechanism.

3 For vehicles fitted with vacuum servo or power operated brakes, the engine must be idling whilst these inspections are conducted.

4 A vacuum servo can be checked for correct operation by applying the brake several times to exhaust the system, then starting the engine with the brake pedal depressed. The brake pedal resistance will alter as the engine starts.

5 On some vehicles a complete inspection of the service brake mechanism requires the bonnet to be opened.

6 These inspections must be carried out without any dismantling. It is therefore accepted that it is not always possible completely to examine all items.

Method of inspection		Principal reasons for failure	Notes
4.1.8	Examine, where practicable and without dismantling, the security of brake back-plates or disc caliper housings.	13 An insecure brake back-plate or disc caliper housing.	
4.1.9	Examine the condition of the chassis or body structure and surrounding panelling in the region of the master cylinder mounting.	14 Excessive corrosion, severe distortion or fracture of the vehicle structure or panelling within 30 cm (12") of the master cylinder mounting.	

5 BRAKE PERFORMANCE TEST

The brake performance test must be carried out on a roller brake tester designated as acceptable for the statutory tests, except in the case of vehicles for which a roller brake tester is not appropriate, or premises where approval has been granted for the test to be carried out using a decelerometer.

Method where the test is to be conducted on a slow speed roller brake tester.

(*NOTE:* The method detailed below refers to roller brake testers under manual control. For roller brake testers under automatic control, follow the general principles given below as modified by the equipment manufacturers' operating instructions).

5.1 Examine the tyres of the vehicle to ensure that they are not obviously under-inflated. (See Note 2).

5.2 Determine whether the vehicle has a split (dual) braking system. (See Note 4).

5.3 Locate the front wheels of the vehicle in the rollers of the brake tester and then run both sets of rollers together to align the vehicle.

5.3.1 With one set of rollers revolving at a time, gradually depress the service brake (foot brake) to determine the maximum braking effort at each front wheel. When the

BRAKE EFFICIENCY AND BALANCE (See Note 6).

A For vehicles having four wheels with the service (foot) brake operating on all four wheels and the parking (hand) brake operating on at least two wheels, and which do NOT have a split (dual) braking system.

Notes

1 The use of a roller brake tester may not be appropriate on vehicles with a permanently engaged four wheel drive, with a belt driven transmission, or fitted with brakes where the servo operates only when the vehicle is moving.

2 The testing of vehicles fitted with ice studded tyres will damage the brake tester roller friction surface. It is advisable to ensure before the roller brake test that the tyres are not damaged and are free from stones embedded in the tread.

3 Vehicles having automatic transmission must never be roller brake tested with the gear selector in the "P" park position.

4 To determine whether the vehicle has a split or dual braking system, check the number of hydraulic pipes leading from the master cylinder. A split or dual system normally has two pipes, or two separate master cylinders.

Method of Inspection	Principal reasons for failure	Notes
maximum braking effort has been determined, release the service brake.	1 The service brake efficiency is less than 50%. (See Note 5).	5 *Occasions will arise when the required brake efficiency is just obtained or just exceeded, but the tester knows that a higher performance figure is normally obtainable for the type of vehicle. In such cases although the vehicle has passed the brake performance test, the tester should advise, on the Inspection Report (VT30), that the braking system appears to be in need of adjustment or repair.*
5.3.2 Having noted these maximum braking efforts, start both sets of rollers, then gradually depress the service brake and watch the way in which the braking effort at each wheel builds up. Gradually release the service brake and watch the way in which the braking effort at each wheel reduces. Note the out-of-balance in braking effort between wheels on either side of the vehicle.	2 The parking brake efficiency is less than 25%. (See Notes 5 and 8). 3 When the out-of-balance of the brakes on the steered roadwheels is such that the lower braking effort is not at least 75% of the higher braking effort. 4 A low braking effort is recorded from the brake on any roadwheel, indicating clearly that the brake on that wheel is not functioning correctly.	6 *The efficiency figures given in this subsection apply to four wheeled vehicles having a service brake (foot brake) operating on four wheels and a parking brake (handbrake) operating on at least two wheels. For information on other vehicles such as three wheeled and veteran vehicles see Appendix D Section F.*
5.3.3 If the vehicle has a parking brake (handbrake), which operates on the front wheels, repeat the process outlined in para 5.3.1 above, using this brake and keeping the "hold-on" button or trigger depressed the whole time.	5 A braking effort recorded at a roadwheel, even though the brake is not applied, indicating that a brake is binding.	7 *The maximum measurable braking occurs just before slipping takes place between the tyre and rollers of the brake test machine.*
5.4 Having released the brakes, drive the vehicle forward until the rear wheels are located in the rollers, running them together as in 5.3 to align the vehicle.	6 Evidence of severe brake grabbing or judder as the brake is applied. 7 The braking efforts at the roadwheels do not increase at approximately the same rate when the service brake is applied gradually.	8 *For some types of vehicles (particularly with front wheel drive) it may be necessary to load the rear of the vehicle in order to increase the adhesion between tyres and rollers.*
5.4.1 With one set of rollers revolving at a time, gradually depress the service (foot) brake to determine the maximum braking effort of each rear wheel.	8 The braking efforts at the roadwheels do not decrease at approximately the same rate when the service brake is released gradually.	9 *For vehicles with servo assisted or power braking systems, the engine must be running (idling) when the service brake is tested.*
5.4.2 Having noted these maximum braking efforts, release the service brake and then start both sets of rollers. Again gradually depress the service brake and watch the way in which the braking effort builds up. Gradually release the service brake and watch the way in which the braking effort at each wheel reduces.		10 *Some vehicles have a trailing arm type rear suspension design. When the front wheels of these vehicles are tested at the same time, the reaction of the rear trailing arms allows the vehicle to move rearwards under the action of the braked front wheels on the rollers. This may allow the front wheels to disengage sufficiently from the rollers to allow the tripping mechanism to operate prematurely and to give a false*

Method of inspection	Principal reasons for failure	Notes
5.4.3 If the vehicle has a parking brake (handbrake) which operates on the rear wheels, repeat the process as outlined in para 5.3.1, using this brake and keeping the "hold-on" button or trigger depressed the whole time.	B For vehicles having four wheels with the service (foot) brake operating on all four wheels and the parking (hand) brake operating on at least two wheels, and which DO have a split (dual) braking system.	*reading of braking effort. It may, therefore, be necessary to test the performance of the front brakes of these vehicles separately. The brake of the non-revolving wheel will then hold the vehicle stationary.*
	1 The service brake efficiency is less than 50%. (See Note 5).	*11 When a veteran car or a vehicle with special controls is submitted for test, the driver of the vehicle should be allowed to drive during the test, if he so wishes.*
	2 The parking brake efficiency is less than 16%. (See Notes 5 and 8).	*12 In some cases it may be necessary to chock the roadwheels of the vehicle during a roller brake test.*
	3 When the out-of-balance of the brakes on the steered roadwheels is such that the lower braking effort is not at least ¾ of the higher braking effort.	
	4 A low braking effort is recorded from the brake on any roadwheel, indicating clearly that the brake on that wheel is not functioning correctly.	
	5 A braking effort recorded at a roadwheel, even through the brake is not applied, indicating a brake is binding.	
	6 Evidence of brake grabbing or judder as the brake is applied.	
	7 The braking efforts at the roadwheels do not increase at approximately the same rate when the service brake is applied gradually.	
	8 The braking efforts at the roadwheels do not decrease at approximately the same rate when the service brake is released gradually.	

Method of inspection | Principal reasons for failure | Notes

6 METHOD OF CALCULATING BRAKE PERFORMANCE

6.1 Total up the braking effort recorded from all the wheels of the vehicle when the service brake is applied. Total up the braking effort recorded from the appropriate wheels when the parking brake is applied.

6.2 From the data provided for the weight of the vehicle, which must contain an element of 140 kg or 300 lbs, as an allowance for the weight of the driver, fuel and tools, calculate the following.

6.2.1 The percentage brake efficiency of the service brake. This is the total braking effort when the service brake is applied, divided by the weight of the vehicle multiplied by 100. (See 5.3.1 and 5.4.1).

6.2.2 The percentage braking efficiency of the parking brake. This is the total braking effort when the parking brake is applied divided by the weight of the vehicle multiplied by 100. (See 5.3.3 and 5.4.3).

6.2.3 The out-of-balance of the braking effort on the steered wheels when the service brake is applied. This is obtained by comparing the maximum braking effort at each steered wheel when the service brake is applied to both wheels at the same time. (See 5.3.2).

Method of inspection	Principal reasons for failure	Notes

7 BRAKE PERFORMANCE TEST

7.1 Decelerometer Test

7.1.1 Using a decelerometer of a type designated as acceptable for the statutory test, set up the meter inside the vehicle in accordance with the manufacturer's instructions. Drive the vehicle at a steady speed of not more than 32 kph (20 mph) (See Note 1) and note the brake efficiency recorded, firstly when the service (foot) brake and secondly when the parking (hand) brake is applied smoothly and progressively. The brakes must not be jabbed or snatched otherwise an inaccurate reading will be obtained.

7.1.2 While the vehicle is decelerating under the action of the brakes, notice whether there is a tendency for the steering wheel to pull or the vehicle to swerve.

Principal reasons for failure

BRAKE EFFICIENCY AND BALANCE (See Note 6).

A For vehicles having four wheels, with the service (foot) brake operating on all four wheels and the parking (hand) brake operating on at least two wheels and which do NOT have a split (dual) braking system.

1 The service brake efficiency is less than 50%. (See Note 5).

2 The parking brake efficiency is less than 25%. (See Notes 5 and 8).

3 When the brakes are applied there is a severe pull to one side on the steering wheel, or the vehicle swerves appreciably from the direction of travel.

B For vehicles having four wheels, with the service (foot) brake operating on all four wheels and the parking (hand) brake operating on at least two wheels and which DO have a split (dual) braking system.

1 The service brake efficiency is less than 50%. (See Note 5).

2 The parking brake efficiency is less than 16%. (See Notes 5 and 8).

3 When the brakes are applied there is a severe pull to one side on the steering wheel, or the vehicle swerves appreciably from the direction of travel.

Notes

1 *The requirement for a steady road speed during a brake test by decelerometer means that the vehicle must always be driven on a road having a good surface, suitable for brake tests when dry or wet, and having the minimum of traffic. A particular public road should not be used for tests to an extent which would justify complaints from local residents.*

2 *For Notes 5, 6 and 8 see page 30.*

Method of inspection	Principal reasons for failure	Notes
8 BRAKE PERFORMANCE TEST		
8.1 High Speed Roller Brake Tester		
8.1.1 Locate the vehicle on the high speed tester and carry out the brake performance test in accordance with the equipment manufacturers' instructions and determine:	A For vehicles having four wheels, with the service (foot) brake operating on all four wheels and the parking (hand) brake operating on at least two wheels and which do NOT have a split (dual) braking system. (See Note 6).	*1 With most types of high speed roller tester, in order to assess the out-of-balance, it will be necessary to drive the vehicle on a road and apply the service brake and determine whether the vehicle swerves appreciably from the direction of travel or there is a severe pull to one side on the steering wheel.*
8.1.2 The brake efficiency of the service brake,	1 The service brake efficiency is less than 50%. (See Note 5).	*2 For Notes 5, 6 and 8 see page 30.*
8.1.3 The brake efficiency of the parking brake,		
8.1.4 The brake balance of the steered wheels (See Note 1).	2 The parking brake efficiency is less than 25%. (See Notes 5 and 8).	
8.1.5 That the brake effort at each wheel is such that it is clear that the brake is working satisfactorily.	3 A low braking effort is recorded from the brake on any roadwheel, indicating clearly that the brake on that wheel is not functioning correctly.	
8.1.6 That the brake on each wheel is not grabbing.	B For vehicles having four wheels, with the service (foot) brake operating on all four wheels and the parking (hand) brake operating on at least two wheels and which DO have a split braking system.	
	1 The service brake efficiency is less than 50%. (See Note 5).	
	2 The parking brake efficiency is less than 16%. (See Notes 5 and 8).	
	3 A low braking effort is recorded from the brake on any roadwheel indicating clearly that the brake on that wheel is not functioning correctly.	

Section IV

Tyres (Including Roadwheels)

Method of inspection		Principal reasons for failure		Notes	
1 TYRES					
1.1	Note the type of structure eg bias-belted, cross-ply or radial ply and nominal size of the tyre fitted to each wheel. (See Note 1).	1	One tyre is of a different nominal size or structure type from the other on the same "axle". (See Notes 2 and 6).	*1*	*The inspection does not include spare tyres: however, if a defect is seen, advise the owner on the inspection form.*
1.2	Examine each tyre for: —cuts —lumps, bulges or tears —separation of the tread —exposure of the ply or cord —incorrect seating in the wheel rim —under inflation (See Note 4) —valve condition and alignment.	2	One tyre on a twin wheel is of a different structure type or nominal diameter from that of its twin (See Note 6).	*2*	*Special light-weight or space-saving wheels and tyres fitted as a road wheel are to be regarded as a reason for failure.*
1.3	Check the tread pattern over the whole breadth of the complete circumference of the tyre. Check also that the tread depth meets the requirements using a depth gauge accepted for MOT testing.	3	A vehicle with radial-ply tyres fitted to the front wheels and cross-ply or bias-belted tyres fitted to the rear wheels. (See Notes 2, 6 and 7).	*3*	*It is not possible to see every part of a tyre, especially the tread contact area when twin wheels are fitted or the body shrouds the tyres. The vehicle must be moved to expose the hidden parts of tyres and the examination completed from under the vehicle if necessary.*
1.4	Check for a tyre fouling on any part of the vehicle.	4	A tyre with a re-cut tread. (See Note 12).	*4*	*Under-inflation of tyres is not a reason for failure. However, it may not be advisable to conduct a brake test because of possible damage to the tyre or a headlight test if alignment is likely to be affected. (Advise the owner on the inspection form.)* *A "punctured run-flat" tyre is a reason for failure.*
1.5	Check tyres on twin wheels for wall contact.	5	A tyre with a cut which reaches the ply or cord. (See Note 8).	*5*	*Tyres which appear to be of inadequate size, ply or speed rating for the vehicle or its use are to be noted as unsuitable in the "Remarks" column of the inspection form.*
		6	A tyre with a lump, bulge or tear caused by separation or partial failure of its structure (this includes any lifting of the tread). (See Note 9).	*6*	*Steel and fabric radial-ply tyres are to be regarded as the same structure type.*
		7	A tyre fouling on any part of the vehicle.		
		8	A tyre with the tread pattern not visible over the whole tread area and the depth of which is not at least 1 mm throughout a single circumferential band of at least three-quarters of the tread breadth. (See Notes 10 and 11).		

Method of inspection	Principal reasons for failure	Notes
	9 Tyres on twin wheels making side-wall contact when standing on a flat surface and correctly inflated.	*7 This requirement does not apply to vehicles fitted with twin wheels.*
	10 A valve badly deteriorated or misaligned.	*8 Superficial scuffing or cutting of the tyre, not deep enough to expose the ply or cord, are not reasons for failure.*
		9 Care should be taken to distinguish on radial-ply tyres, between normal undulations in the carcass resulting from manufacturing and lumps or bulges caused by structural deterioration.
		10 "Breadth of tread" means that part of the tyre which can contact the road under normal conditions of use measured at 90° to the peripheral line of the tread.
		11 Tyres which do not meet the three-quarters width rule when new must have a minimum of 1 mm tread depth over the whole of the original tread pattern.
		12 Re-cut tyres are permitted to be fitted to: – vehicles of unladen weight exceeding 3050 kg or those having an unladen weight exceeding 2540 kg which have been adapted to carry more than 7 passengers (not including the driver).

2 ROADWHEELS

Method of inspection	Principal reasons for failure	Notes
2.1 Carry out a visual inspection of all wheels except the spare for obvious signs of damage, particularly of the bead rim. This inspection is to be carried out without removing hub caps or wheel trims.	1 A wheel which is badly damaged, distorted or cracked.	*The condition of the spare wheel is not included in the inspection: however, if a defect is seen, advise the owner on the inspection form.*
	2 A wheel which has a bead rim which is badly distorted.	
2.2 Check that the wheel is securely attached.	3 A wheel which is insecure.	
2.3 If visible, check wheel attachments.	4 Loose or missing wheel nuts, studs or bolts.	

Section V

Seat Belts

Method of inspection		Principal reasons for failure		Notes	
1 SEAT BELTS					
1.1	See that the driving seat and the appropriate front passenger's seat (see notes) are each provided with a seat belt of the type which constrains the upper part of the seat occupant's body.	1	The driving seat and front passenger's seat are not provided with a seat belt which constrains the upper part of the seat occupant's body.	*1*	*Vehicles required to have seat belts fitted are:* *(i) Three-wheeled vehicles over 408 kgs (8 cwt) unladen registered on or after 1 January 1965 and which were constructed after 30 June 1964,* *(ii) Three-wheeled vehicles over 254 kgs (5 cwt) unladen registered on or after 1 September 1970 and which were constructed after 1 March 1970,* *(iii) Motor cars first registered on or after 1 January 1965 and which were constructed after 30 June 1964 and* *(iv) Goods vehicles not over 1525 kgs (30 cwt) unladen first registered on or after 1 April 1967 and which were constructed after 1 September 1966.*
1.2	Pull each seat belt webbing against its anchorage and see that it is properly secured to the vehicle structure.	2	A seat belt which is not securely fixed to the seat or to the structure of the vehicle. For example, one or more fixing bolts not secure.		
		3	For seats with integral seat belts, any insecure attachment of the driving seat or front passenger seat to the vehicle structure.		
1.3	Examine the condition of each seat belt webbing for cuts or obvious signs of deterioration.	4	A cut or serious deterioration in any part of the seat belt webbing.		
1.4	Fasten each belt locking mechanism and then try to pull the locked sections apart. Operate the mechanism, whilst pulling on the belt to determine that the mechanism releases when required.	5	A locking mechanism of a seat belt which does not secure or release the belt as intended when the webbing is pulled.	*2*	*For seats with integral seat belts it may not be possible to carry out an examination of the fixing of the seat belt to the seat.*
				3	*Only the seat belts provided for the driving seat and front passenger seats are included in the*

Method of Inspection	Principal reasons for failure	Notes
1.5 Examine the condition of the attachment fittings and adjusting fitting on each belt for distortion or fracture.	6 An attachment fitting or adjustment fitting of a seat belt which is fractured or badly deteriorated.	*inspection. If there is more than one front passenger's seat (eg a three seater bench seat) then the passenger seat furthest from the driving seat must be fitted with a seat belt. This is the seat belt included in the inspection.*
1.6 As far as is practicable without dismantling, check the condition of the vehicle structure in the vicinity of the seat belt anchorage points. The condition of floor mounted anchorage points may best be inspected from underneath the vehicle.	7 Excessive corrosion, serious distortion or a fracture in any load bearing member of the vehicle structure or panelling within 30 cm (12") of a seat belt anchorage.	
1.7 If the seat belt is of the retracting type, pull a section of the webbing from the retracting unit and then release it to see that the webbing is automatically wound in to the retracting unit. Manual assistance may be required with some types of retracting seat belts before retraction takes place.	8 The retracting mechanism of a seat belt has failed to such an extent that no winding in of the webbing occurs.	

Section VI

General Items (Windscreen Wipers and Washers, Exhaust Systems and Horns)

Method of inspection	Principal reasons for failure	Notes
1 FUNCTIONING OF WINDSCREEN WASHERS		
1.1 Operate the windscreen washer mechanism and note that liquid is emitted from the washer.	1 The windscreen washer intended to wash the windscreen on the driver's side does not function. 2 The windscreen washer intended to wash the driver's side of the windscreen, emits liquid but it is evident that the flow is such that it will not fulfil the function of cleaning the windscreen to afford the driver an adequate view to the front and forward of the near and off sides of the vehicle.	*1 Every vehicle without an opening windscreen, is required to be fitted with one or more windscreen wipers and is also required to have a windscreen washer capable of clearing the windscreen in conjunction with the windscreen wiper(s).* *2 There is no requirement as to the number of jets to be provided for the washers.*
2 FUNCTIONING OF WINDSCREEN WIPERS		
2.1 Operate the windscreen wipers and note that they move over an adequate area of the windscreen.	1 A windscreen wiper does not operate over a sufficient area of the windscreen to give the driver an adequate view of the road in front and forward of the near and off sides of the vehicle. 2 A windscreen wiper which has a blade deteriorated to such an extent that the screen is not cleared effectively enough to give the driver an adequate view of the road in front and forward of the near and off sides of the vehicle.	

Method of inspection		Principal reasons for failure		Notes
3 EXHAUST SYSTEM				
3.1	Assess subjectively the effectiveness of the silencer in reducing exhaust noise to a level considered to be average for the vehicle.	1	A silencer which is in such a condition, or of such a type, that the noise emitted from the vehicle is clearly unreasonably above the level to be expected from a similar vehicle with a silencer in average condition.	*It is usually necessary to open the bonnet to carry out a complete inspection of the exhaust system.*
3.2	Examine the exhaust system for signs of exhaust gas leaks, paying particular attention to joints and any parts where the system has been repaired.	2	A major leak of exhaust gas from the exhaust system. A repair to an exhaust system which effectively prevents leaks is acceptable, providing the system is structurally sound.	
3.3	Examine the condition of the silencer outer casing for serious deterioration and the exhaust system for completeness.	3	A missing silencer.	
3.4	Examine the condition of the mountings of the exhaust system to the vehicle.	4	A missing tail pipe.	
		5	An exhaust system mounting missing, or one which is in such a condition that it does not fully support the exhaust system.	
4 FUNCTION OF AUDIBLE WARNING DEVICE				
4.1	Operate the horn and listen for the audible warning.	1	The horn control is missing.	
		2	The horn does not function.	
		3	The horn control is not accessible to the driver.	
		4	The horn operates but the sound from it has such a low volume that it is unlikely to be audible to another road user.	
		5	The audible warning is a gong, bell, siren or two tone horn and the vehicle is not one which is permitted by Regulation to have any of these items.	

Method of inspection		Principal reasons for failure		Notes	
5	**CONDITION OF THE VEHICLE STRUCTURE**				
5.1	With the vehicle over a pit or on a raised hoist examine the vehicle structure for any fractures, damage or corrosion, not within the prescribed areas, which is likely prejudicially to affect the correct functioning of the braking system or the steering gear.	1	Any fracture, damage or corrosion in the vehicle structure of such a serious nature as to affect prejudicially the correct functioning of the braking system or steering gear.	*1*	*The prescribed areas are those specifically detailed in the manual.*

Appendix A

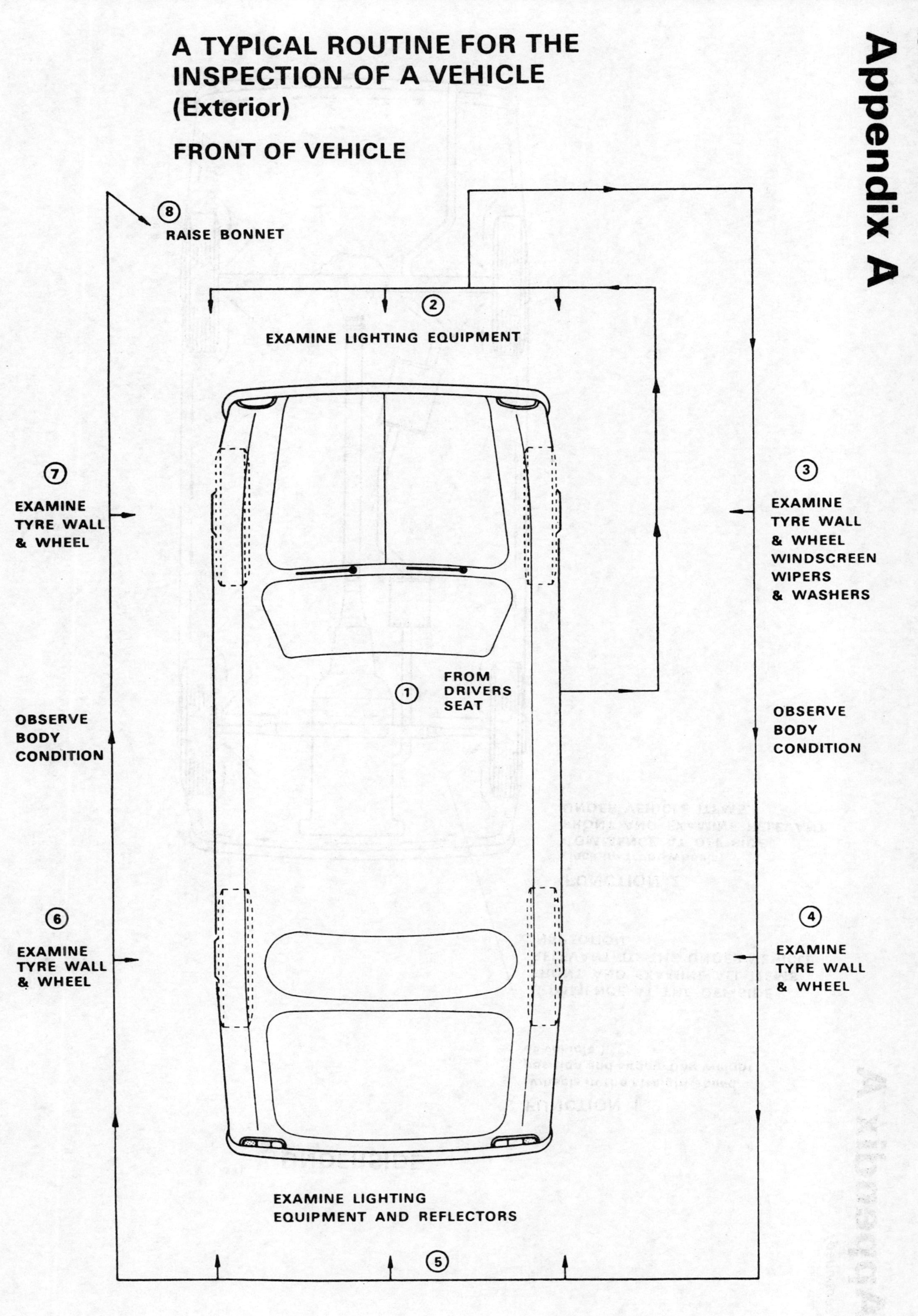

A TYPICAL ROUTINE FOR THE INSPECTION OF A VEHICLE (Exterior)
FRONT OF VEHICLE
8 RAISE BONNET
2 EXAMINE LIGHTING EQUIPMENT
7 EXAMINE TYRE WALL & WHEEL
3 EXAMINE TYRE WALL & WHEEL WINDSCREEN WIPERS & WASHERS
1 FROM DRIVERS SEAT
OBSERVE BODY CONDITION
OBSERVE BODY CONDITION
6 EXAMINE TYRE WALL & WHEEL
4 EXAMINE TYRE WALL & WHEEL
EXAMINE LIGHTING EQUIPMENT AND REFLECTORS
5

Appendix A

(continued)

UNDERSIDE

FUNCTION 1
(wheels in the straight ahead position and supporting weight of vehicle.)

COMMENCE AT THE OFF-SIDE FRONT AND EXAMINE ALL ITEMS RELEVANT TO THE UNDER-VEHICLE INSPECTION.

FUNCTION 2
(jack up front wheels)
COMMENCE AT OFF-SIDE FRONT AND EXAMINE RELEVANT UNDER-VEHICLE ITEMS.

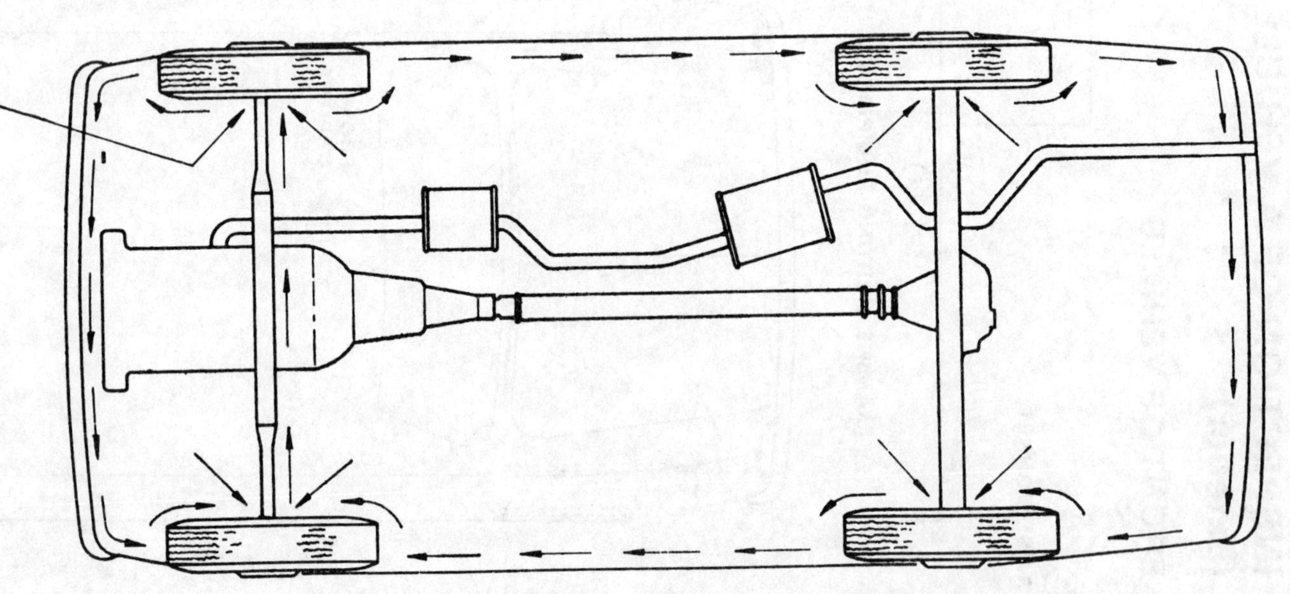

Appendix B

Department of Transport

Road Traffic Act 1972 section 43

MOT Inspection Report

FILE COPY

Serial No.00000001AA

Vehicle Reg. Mark ____________ VIN or Chassis no. ____________ Year of manufacture ____________

Make & Model ____________ Colour ____________ Recorded mileage ____________

A Testable Item	Testers Manual Ref. (see over)	Pass	Fail	Reasons for Failure and Remarks
Section I — Lighting Equipment				
Front Lamps	I/1			
Rear Lamps	I/1			
Headlamps	I/2			
Headlamp Aim	I/6			
Stop Lamps	I/3			
Rear Reflectors	I/4			
Direction Indicators	I/5			
Section II — Steering & Suspension				
Steering Controls	II/1			
Steering Mechanism	II/2			
Power Steering	II/3			
Transmission Shafts	II/2,4,9			
Stub Axle Assemblies	II/5			
Wheel Bearings	II/4			
Suspension	II/5,6,7,8,9			
Shock Absorbers	II/10			
Section III — Brakes				
Service Brake Condition	III/3,4			
Parking Brake Condition	III/1,2			
Service Brake Performance	III/5,6,7,8			
Parking Brake Performance	III/5,6,7,8			
Service Brake Balance	III/5,6,7,8			
Section IV — Tyres & Wheels				
Tyre Type	IV/1			
Tyre Condition	IV/1			
Roadwheels	IV/2			
Section V — Seat Belts				
Security of Mountings	V/1			
Condition	V/1			
Operation	V/1			
Section VI — General Items				
Windscreen Washers	VI/1			
Windscreen Wipers	VI/2			
Horn	VI/4			
Exhaust System	VI/3			
Silencer	VI/3			
Vehicle Structure	VI/5			

B **Test Result** 1 ☐ PASS Test certificate issued. No: ____________

2 ☐ FAIL see below

C **Notice of Refusal of a Test Certificate** (see notes overleaf)

1 ☐ For the reasons shown in the above Inspection Report

2 ☐ Because the inspection could not be completed, for the following reasons:

D **Warning** In my opinion the vehicle is DANGEROUS to drive because of the following defects:

Authentication Stamp

E Signed ____________ Date ____________
(Tester/Inspector)

Name (block capitals) ____________ Testing Station no. ____________

YOU ARE ADVISED TO KEEP THIS FORM

VT30

Appendix C
Assessing Corrosion

The effect of corrosion on the safety of a vehicle is a difficult matter to resolve, since it depends not only on the extent of the corrosion, but also on the function of the section on which it has occurred. Quite a small amount of corrosion of an important part of a vehicle structure can render the vehicle unsafe where it destroys the continuity of the load bearing structure, whereas large amounts of corrosion of unimportant sections may have no effect on its safety. Similarly corrosion of a particular part, such as body sills, may be very important on one type of construction, but of less importance on another type. This is shown in figures A to D where the shaded portions indicate the important load bearing parts of different typical vehicle constructions.

In order to assist with the assessment of corrosion, the Manual identifies those parts of the vehicle structure located near to particular components which are important and to which particular attention has to be paid during an inspection.

Having identified the important load bearing members on a vehicle, the tester should then attempt to determine the extent of any corrosion on them. He should do this by pressing hard against them and noting the amount of "give" or disintegration which results from this. It should not be necessary to use a sharp instrument to "dig" at the structure nor to subject it to heavy impact blows in order to establish the extent of corrosion. Often, however, it is necessary to tap the component lightly and to listen for the difference in sound which results from unaffected metal, compared with that from badly corroded metal, or metal which has been treated with filler to camouflage the corrosion. Excessively corroded metal, or metal treated with a filler, emits a duller sound than does unaffected metal.

Having determined the extent of corrosion, the tester then has to decide whether it has materially affected the strength of the component concerned. This is usually a matter for the tester using his experience and judging the extent of the corroded area relative to the total area of the component and the amount of sound metal surrounding the affected area.

If corrosion has not materially reduced the metal thickness to weaken it (eg there is only surface rust), clearly the tester should not regard this as rendering the component defective, although it would be helpful for him to inform the vehicle owner of the fact that corrosion has started. On the other hand, if there are holes in a critical component or large areas are no longer rigid or sections are missing and the strength of the component is obviously well below that of its original state, it may be necessary to refuse to carry out a brake test.

When judging corrosion on a vehicle, a good criterion for a tester to adopt is for him to consider whether he personally would feel safe riding at speed in the particular vehicle, with the possibility of an emergency brake stop taking place. If he would feel safe, he should pass the component. If on the other hand he considers that he would feel unsafe, he would obviously be justified in regarding it as defective.

When a corroded area has been repaired, it is essential that this should have been executed properly. The repair should be effected so that any plating or welding extends to a sound part of the load bearing component and should be such that it is clear that the strength of the repair is approximately the same as that of the component originally. In some cases it may be that the only solution to the presence of a corroded load bearing member is for it to be replaced. Repairs to load bearing members or sections by pop-riveting or glass fibre are not acceptable, but these methods may be used for non-load bearing members in some cases.

Appendix C

(continued)

FIGURE 'A'

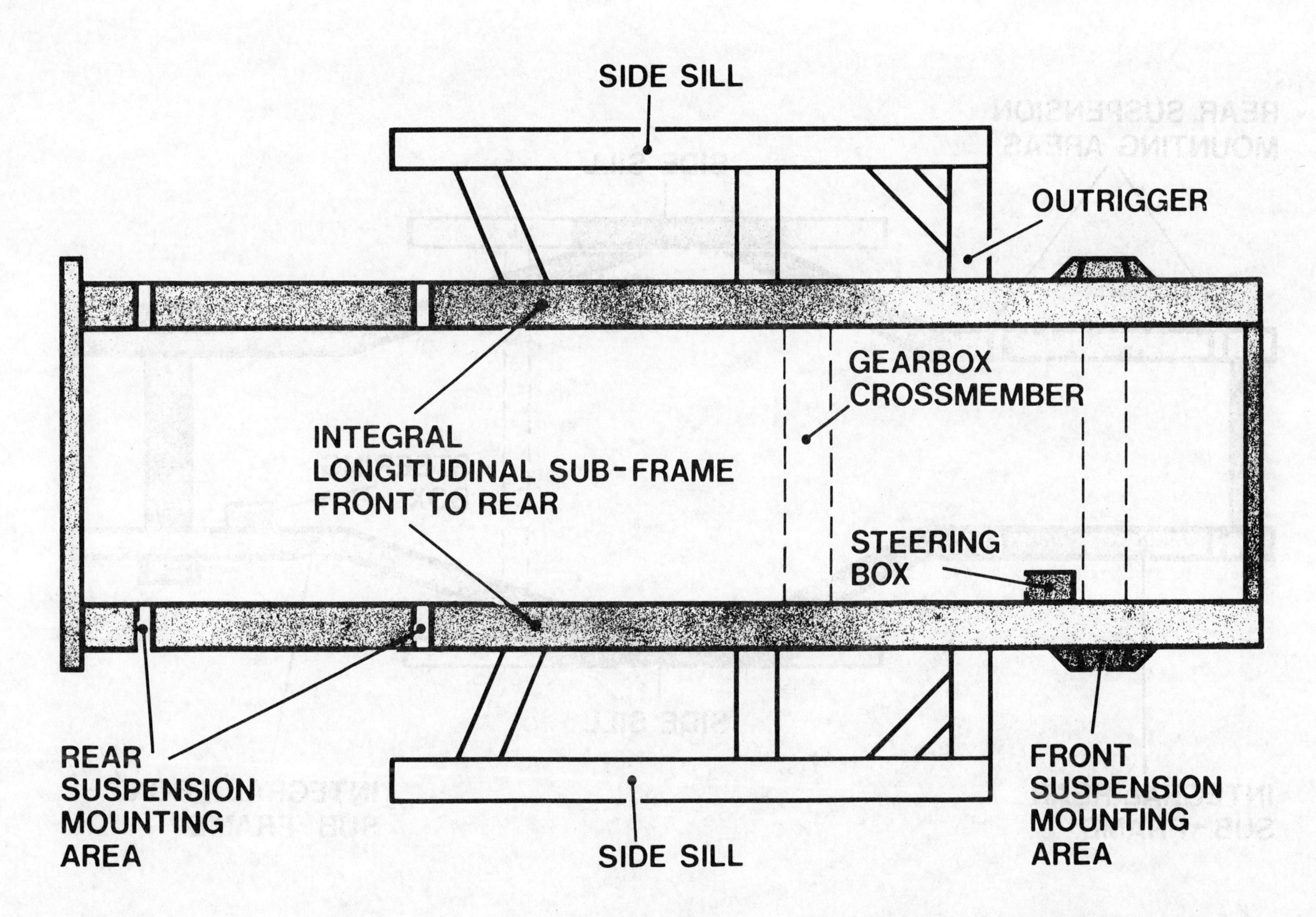

Appendix C

(continued)

FIGURE 'B'

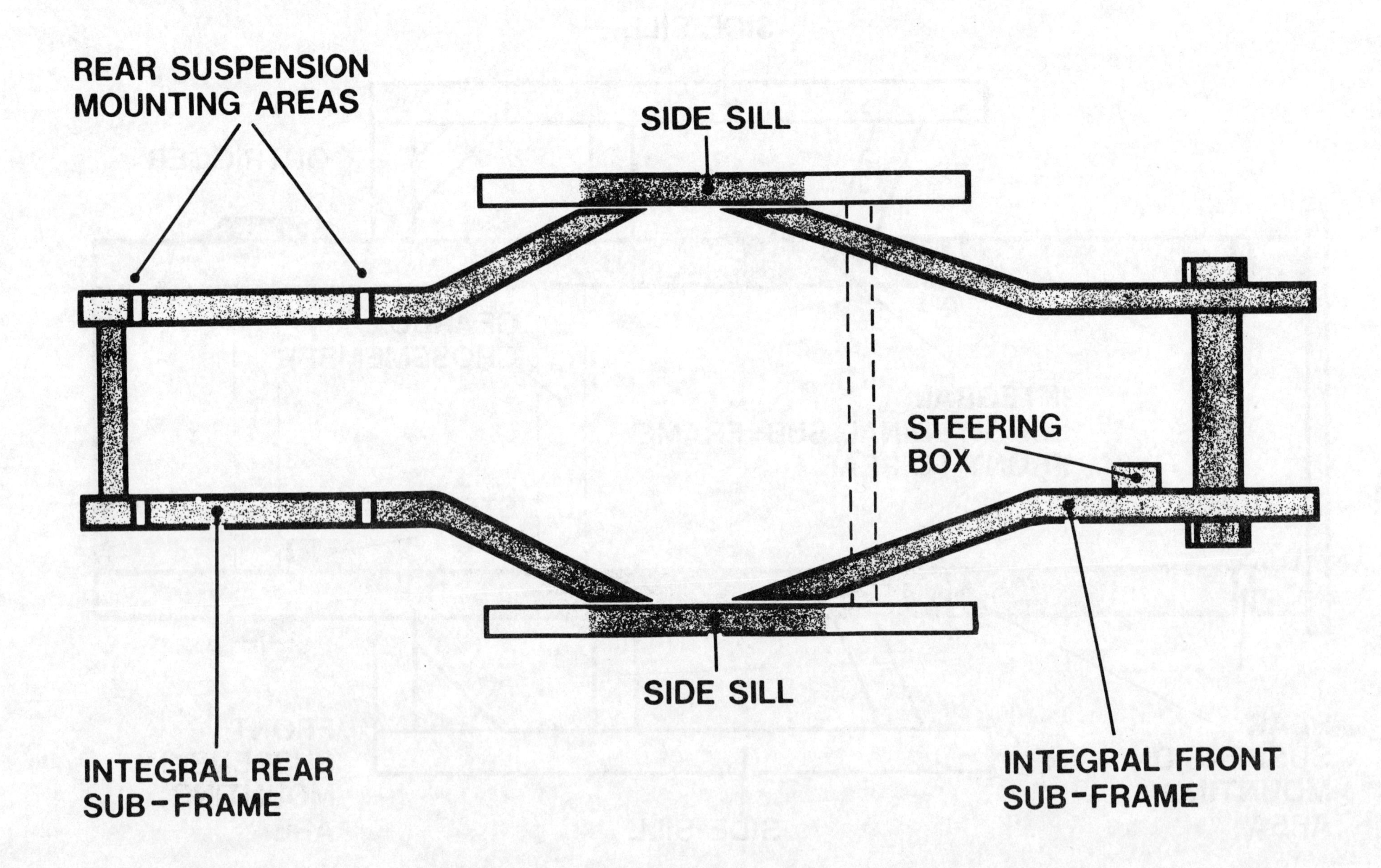

Appendix C

(continued)

FIGURE 'C'

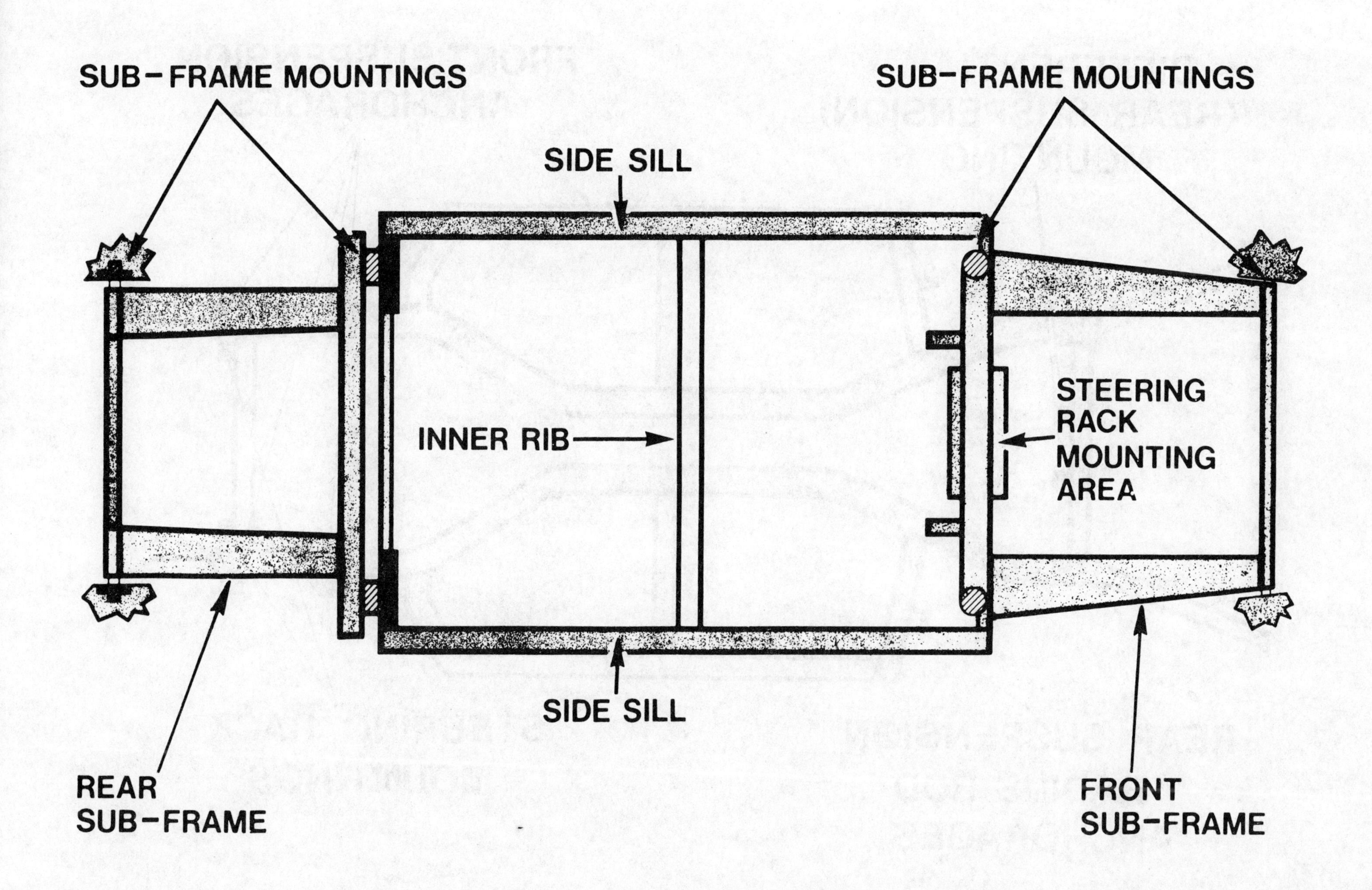

Appendix C

(continued)

FIGURE 'D'

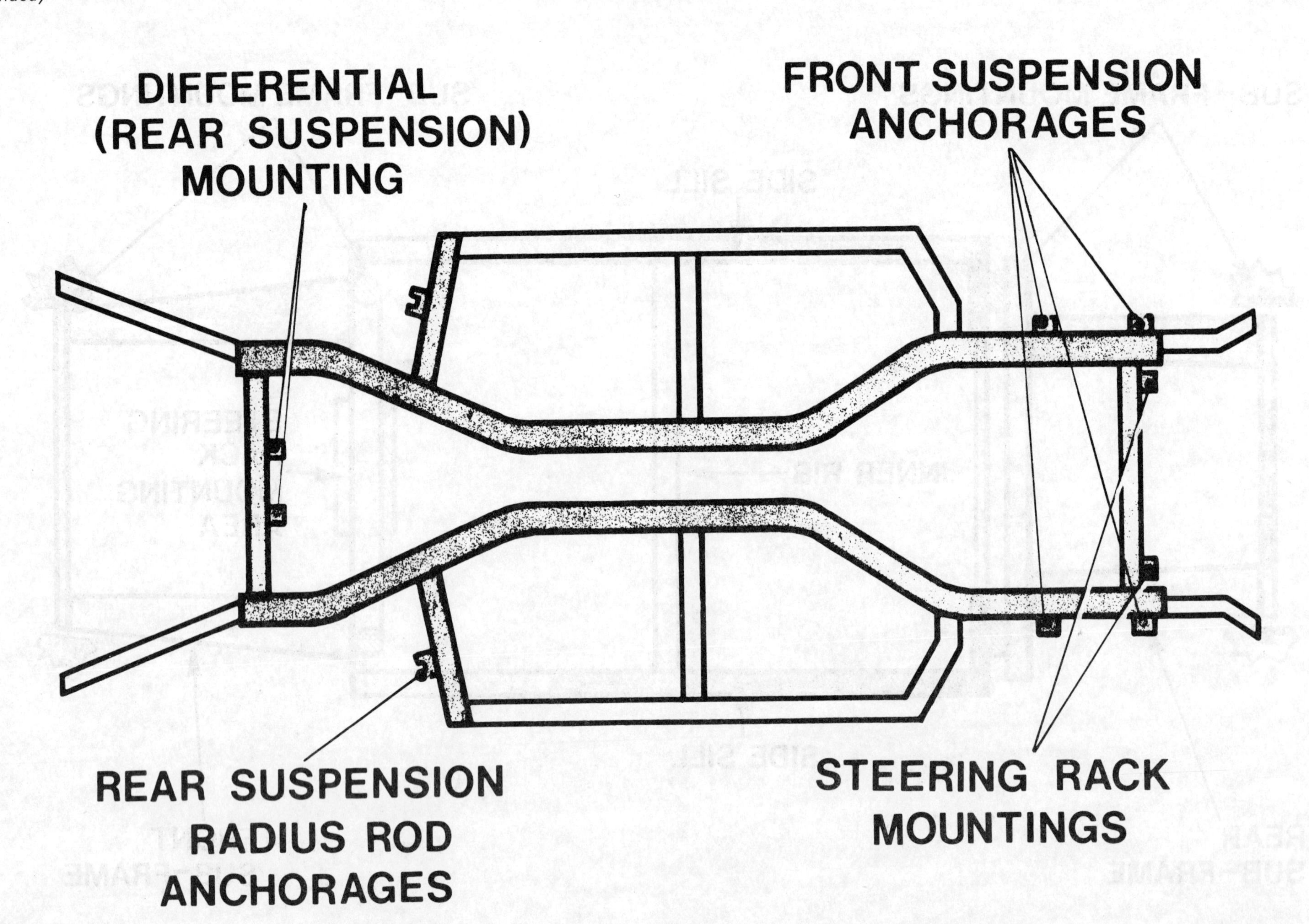

Appendix D

The Prescribed Statutory Requirements for vehicles of Classes III, IV, and V

(ie three and four-wheeled motor cars, heavy motor cars, dual-purpose vehicles (see Appendix E) and light goods vehicles not exceeding 1525 kg (30 cwt) unladen weight and privately licensed passenger-carrying vehicles).

Note: Act and regulation numbers are set in italics at the end of each relevant section or paragraph.

A GENERAL

1 The statutory requirements prescribed by the Motor Vehicles (Tests) Regulations which apply to vehicles submitted for testing are contained in:

The Road Traffic Act 1972 (referred to below as "RTA");

The Motor Vehicles (Construction and Use) Regulations 1973 (referred to below as "C&U"); and

The Road Vehicles Lighting Regulations 1971 (referred to below as "RVLR").

Note: (a) The Motor Vehicles (Construction and Use) Regulations 1973 has been superseded by the 1986 edition.

(b) The Road Vehicles Lighting Regulations 1971 has been superseded by the 1984 edition.

2 This Appendix summarises the regulations upon which this manual is based.

THE STATUTORY REQUIREMENTS

B LIGHTING EQUIPMENT

1 General

Every motor vehicle (with the exception of those in (a) and (b) below) is required to have lighting equipment and rear reflectors to enable it to be used on roads during the hours of darkness and the equipment and reflectors must be maintained in clean and efficient working order at all times whilst the vehicle is in use on roads.

Exceptions:

(a) *If the vehicle has no front or rear lamp at all; and*

(b) *If a front or rear lamp is provided, it is painted over or masked so as to be incapable of immediate use, or if it has no wiring system to connect it to a source of electricity.* *C&U 38 & 97*

2 Obligatory Front and Rear Lamps and Rear Reflectors

Front and rear lamps are normally the "sidelamps" of a vehicle, but the sidelamps may be contained in the headlamp assembly.

Except as indicated in paragraphs 1 and 4 of this section, all motor vehicles required to carry lighting equipment must have:

(a) Two obligatory front lamps which can be used to show a white light to the front visible from a reasonable distance, fixed on opposite sides of the vehicle at the same height from the ground and having diffusing lenses,
RTA Sec 68(1), RVLR 5 & 9(1) (b)

(b) Two obligatory rear lamps which can be used to show a red light to the rear visible from a reasonable distance, fixed on opposite sides of the vehicle at the same height from the ground and
RTA Sec 68(1), RVLR 23 & Sch 1

(c) Two unobscured and efficient red reflectors each facing to the rear and fixed on opposite sides of the vehicle at the same height from the ground.
RTA Sec 69, RVLR 30 & Sch 2

3 Obligatory Headlamps

Definitions:

A "headlamp" is a lamp on a motor vehicle which is designed, when lit, to illuminate the road in front of the vehicle and which is not a fog lamp.

An "obligatory headlamp" is a headlamp required to be carried by a vehicle.

A "matched pair of headlamps" means a pair of headlamps on a vehicle which are the same height from the ground, with one lamp on each side of the vehicle equidistant from the vertical centre line of the vehicle.

A "fog lamp" is a lamp on a motor vehicle which is fitted to be used primarily in conditions of fog or whilst snow is falling.

A headlamp must be in a clean and efficient condition and where a matched pair of headlamps is fitted, they must be symmetrically located on the vehicle and both show the same colour light, which may be either white or yellow.
RVLR 18(3), 19(3) & 20

Each headlamp must be so constructed, fitted and maintained that, when used on a road during the hours of darkness, its beam:

(a) Would be permanently deflected downwards or downwards and to the left so that it would not dazzle a person standing more than $7\frac{1}{2}$ m (25 ft) from the lamp and whose eye level is not less than 106 cms (42") above the horizontal plane on which the vehicle stands; or

(b) Can be deflected downwards or downwards and to the left by the driver so that it would not dazzle the person mentioned in (a); or

(c) Can be extinguished by the operation of a device which at the same time:

(i) Causes a beam of light which complies with (a) to be emitted from the lamp, or

(ii) Deflects the beam of light from another lamp downwards or downwards and to the left so that it would not dazzle the person mentioned in (a), or

(iii) Brings into or leaves in operation a lamp or lamps (other than the obligatory front lamps) which complies with (a).

Note: A lamp cannot be treated as complying with (a) above unless it is so fixed that its centre is not less than 2 ft from the ground.
RVLR 9

4 *Exceptions to the requirements contained in paragraphs 1 to 3 of this section are as follows:*

(a) *Large passenger carrying vehicles first registered before 1 October 1954 are only required to carry one rear lamp,*

(b) *Motor vehicles first registered before 1 January 1931 are not required to carry obligatory headlamps,*

(c) *Motor vehicles having three wheels and first used before 1 January 1972 are only required to carry one headlamp and*

(d) *Motor vehicles having three wheels and an unladen weight of not more than 408 kg (8 cwt) and a width of not more than 1·3 m (4 ft 4 in) first used on or after 1 January 1972 are only required to carry one headlamp.*

C STOP LAMPS

1 Except as indicated in paragraph 3 of this section, all motor vehicles:

(a) First used on or after 1 January 1936 and before 1 January 1971 must have at least one stop lamp showing a diffused steady red light when in operation and fitted at the rear of the vehicle either on or to the right of the centre of the vehicle (if two stop lamps are fitted they must comply with (b) below);
RVLR 72 & 73

(b) First used on or after 1 January 1971 must have two stop lamps each showing a diffused steady red light when in operation and being symmetrically located on each side of the vehicle at the same height from the ground.
RVLR 74 & Sch 6 Pt II

2 Every stop lamp fitted to a motor vehicle shall be maintained in a clean condition and in good and efficient working order.
RVLR 76

3 *Motor vehicles which are NOT required to have stop lamps fitted include those:*

(a) *First used before 1 January 1936,*

(b) *Which may not lawfully exceed 24 kph (15 mph),*

(c) *Which are incapable of exceeding 24 kph (15 mph) on the level under their own power,*

(d) *Which do not have electrically operated lighting equipment and*

(e) *Which have no lamps at all.*
RVLR Sch 6 Pt I

D DIRECTION INDICATORS

1 Except as indicated in paragraph 8 of this section, all motor vehicles first used on or after 1 January 1936 are required to have direction indicators which must be maintained in good and efficient working order while the vehicle is in use on a road.
RVLR 70 & 76

2 Motor vehicles first used on or after 1 January 1936 and before 1 September 1965:

(a) Which are fitted with electric lighting equipment must have either;

(i) Semaphore type direction indicators showing either a steady or a flashing light or

(ii) Flashing type direction indicators;
RVLR 69(1) (a) & Sch 5 Pts I, II & III

(b) Which are NOT fitted with electric lighting equipment, may have semaphore type direction indicators in the form of a hand presenting a white surface to the front and to the rear of the vehicle—see also paragraph 8 below.
RVLR 69(1) (b) & Sch 5 Pt V

3 Motor vehicles first used on or after 1 September 1965 must have flashing type direction indicators.
RVLR 70(1) (b) 71 (4) & Sch 5 Pt III

4 The illuminated colour of every direction indicator must:

(a) Be amber if showing to both front and rear,

(b) Be amber or white if showing only to the front and

(c) Be amber or red if showing only to the rear.

5 The light shown by all illuminated direction indicators must be diffused.

6 Flashing direction indicators must flash at a rate of not less than 60 nor more than 120 flashes per minute. *RVLR Sch 5 Pt I para 6 & 7*

7 If, when in the driving position, the driver cannot see at least one direction indicator on each side to know when it is in operation, the vehicle must be fitted with a device to give either visual or audible warning of the operation of the indicators.
RVLR Sch 5 Pt I para 8

8 *Motor vehicles which are NOT required to have direction indicators fitted include those:*

(a) *Which have lamps not electrically operated or which carry no lamps at all,*

(b) *First used before 1 January 1936,*

(c) *Which may not lawfully exceed 24 kph (15 mph) and*

(d) *Which are incapable of exceeding 24 kph (15 mph) on the level under their own power.*
RVLR 70(2)

E STEERING

1 All steering gear fitted to any motor vehicle must be maintained in good and efficient working order and be properly adjusted. *C&U 95(1)*

2 Springs must be provided between each wheel and the frame of the vehicle. *C&U 12*

3 The bodywork and suspension of the vehicle must be in such a condition as not to cause or be likely to cause danger to any person in or on the vehicle or on a road through any adverse effect on the steering of the vehicle. *C&U 90(1)*

F BRAKES AND BRAKING SYSTEMS

1 Parking Brakes

1.1 All vehicles must have a braking system so designed and constructed that it can at all times be set so as effectually to prevent two at least, or in the case of a three-wheeled vehicle, one at least, of the wheels from turning when the vehicle is not being driven or is left unattended. *C&U 13(1)*

1.2 Every vehicle first used on or after 1 January 1968 must have a braking system capable, when the vehicle is not being driven or is left unattended, of holding the vehicle stationary on a gradient of 1 in 6.25 (16%) by mechanical means only, ie without the assistance of hydraulic, pneumatic or electric aids.
C&U 13(2)

2 Service Brakes

2.1 Except as specified in paragraph 3, all vehicles must have an efficient braking system with two means of operation or two efficient braking systems each having a separate means of operation or one efficient braking system with one means of operation if that system is a split system.
C&U 54(1), (4) & (5) & 59(1), (4) & (5)

2.2 One means of operation of the braking system or systems of every heavy motor car or motor car first used on or after 1 January 1968 must act on all the wheels of the vehicle.
C&U 54(1), (4) & (5) & 59(1), (4) & (5)

3 A motor car first registered before 1 January 1915 under the Motor Car Act 1903 need have only one efficient braking system. *C&U 59(8) (a)*

4 All Braking Systems

4.1 Every part of every braking system fitted to a vehicle, and its system of operation must be maintained in good and efficient working order and be properly adjusted. *C&U 94(1)*

4.2 The bodywork and suspension of the vehicle must be in such a condition as not to cause or be likely to cause danger to any person in or on the vehicle or on a road through any adverse effect on the braking of the vehicle. *C&U 90(1)*

5 Braking Efficiency

Definition:
"Braking efficiency", in respect of the application of brakes to a motor vehicle at any time, means the maximum braking force capable of being developed by the application of those brakes expressed as a percentage of the weight of the vehicle including any driver and load but excluding passengers carried in the vehicle at that time.

5.1 All braking systems must comply with certain specified requirements in respect of their efficiency. (The effect of these requirements is set out in paragraph 5.2 below).

5.2 When the brakes of a vehicle are applied by one or the other means of operation they must have an efficiency of a percentage not less than is specified below.

Type of vehicle	*Minimum % of braking efficiency required*
(a) Vehicles having four or more wheels and required to have two means of operating brakes:	
(i) If each means of operation applies brakes to at least four wheels—the brakes as applied by one of the means (ie the main means)	50
the brakes as applied by the other means (ie the secondary means)	25
(ii) If only one of the means of operation applies brakes to at least four wheels: the brakes as applied by that means (ie the main means)	50
the brakes as applied by the other means (ie the secondary means)	25
(iii) If only one of the means of operation applies brakes to at least four wheels and the system is a split system: the brakes applied by that means (ie the main means)	50
the brakes as applied by other means (ie the secondary means)	16

NOTE: The 16% is the efficiency necessary to meet the requirement to hold the vehicle on a gradient of 1 in 6.25.

Type of vehicle	*Minimum % of braking efficiency required*
(iv) If neither means of operation applies brakes to at least four wheels: the brakes as applied by one of the means (ie the main means)	30
the brakes as applied by the other means (ie the secondary means)	25

Type of vehicle	*Minimum % of braking efficiency required*
(b) Vehicles having three wheels (not motor cycles with sidecars) required to have two means of operating brakes:	
(i) If each means of operation applies brakes to all three wheels: the brakes as applied by one of the means (ie the main means)	40
the brakes as applied by the other means (ie the secondary means)	25
(ii) If only one means of operation applies brakes to all three wheels: the brakes as applied by that means (ie the main means)	40
the brakes as applied by the other means (ie the secondary means)	25
(iii) If neither means of operation applies brakes to all three wheels: the brakes as applied by one of the means (ie the main means)	30
the brakes as applied by the other means (ie the secondary means)	25
(c) Vehicles not required to have two means of operating brakes:	
(i) Vehicles having four or more wheels and at least one means of operation applying brakes to at least four wheels: the brakes as applied by that means	50
(ii) Vehicles with four or more wheels having no means of operation applying brakes to at least four wheels: the brakes as applied by one of the means	30
(iii) Vehicles having three wheels (not motor cycles with sidecars) having one or more means of operation applying brakes to all three wheels: the brakes as applied by one of the means	40
(iv) Vehicles having three wheels (not motor cycles with sidecars) and having no means of operation applying brakes to all three wheels: the brakes as applied by one of the means	30

NOTE: See paragraph 2 above for the minimum efficiency required in respect of a parking brake on a vehicle having a split braking system.

C&U 54 (4) & (5) & 59 (4) & (5) & Sch 4

G TYRES

NOTE: The Motor Vehicles (Construction and Use) Regulations 1978 as amended are referred to below as "C & U".

1 All road wheels fitted to any motor vehicle must be in such condition as not to cause or be likely to cause danger to any person in or on the vehicle or on the road through any adverse effect on the tyres of the vehicle. C & U 97(1).

2 A motor vehicle fitted with a pneumatic tyre may not be used on the road if:

(a) The tyre is unsuitable for the use to which the vehicle is put or is unsuitable having regard to the types of tyre fitted to its other wheels.

(b) The tyre has a cut longer than 25 mm or 10% of the section width of the tyre, whichever is the greater, and deep enough to reach the ply or cord.

(c) The tyre has any lump, bulge or tear caused by the separation or partial failure of its structure.

(d) The tyre has any portion of the ply or cord exposed.

(e) The base of any groove which showed in the original tread pattern of the tyre is not clearly visible.

(f) either

(i) the grooves of the tread pattern of the tyre do not have a depth of at least 1 mm throughout a continuous band measuring at least three-quarters of the breadth of the tread and round the entire circumference of the tyre, or

(ii) in the case where the original tread pattern of the tyre does not extend beyond three-quarters of the breadth of the tread, the base of any groove which showed in the original tread pattern does not have a depth of at least 1 mm.

(g) The tyre has any defect which might in any way cause damage to the surface of the road or danger to persons on or in the vehicle or to other persons using the road.

C&U 107(1)(a), (c), (d), (e), (f) and (g) and 107(5).

3 Re-cut pneumatic tyres may only be fitted to vehicles of unladen weight exceeding 3050 kg or which have an unladen weight exceeding 2540 kg and have been adapted to carry more than 7 passengers (not including the driver), provided that:–

(a) The ply or cord of the tyre has not been cut or exposed by the re-cutting process, or

(b) The tyre has been wholly or partially re-cut to the manufacturer's re-cut tread pattern, or the original tread pattern as appropriate.

C&U 107(4)

H SEAT BELTS

1 Vehicles to which seat belt requirements apply are:

(a) Motor cars (including three-wheeled vehicles exceeding 408 kg (8 cwt), unladen weight) manufactured on or after 30 June 1964 and registered on or after 1 January 1965,

(b) Goods vehicles not exceeding 1525 kg (30 cwt) unladen weight (excluding electrically propelled goods vehicles) manufactured on or after 1 September 1966 and registered on or after 1 April 1967 and

(c) Three-wheeled vehicles exceeding 254 kg (5 cwt) unladen weight manufactured on or after 1 March 1970 and first used on or after 1 September 1970. *C&U 17(1) & (2)*

2 All vehicles to which the seat belt requirements apply must have seat belts and suitable anchorage points provided for the driver's seat and for the foremost forward-facing passenger seat which is furthest from the driver. *C&U 17(3) & (4)*

3 All obligatory seat belts, seat belt mechanisms, anchorage points and load-bearing members of the vehicle structure within 30 cms (12") of an anchorage point must be in such a condition as not to cause or be likely to cause danger to any person in or on the vehicle or on a road. *C&U 90(1)*

I GENERAL ITEMS

1 Windscreen Wipers

1.1 Unless an adequate view can be obtained to the front of the vehicle without looking through the windscreen (by opening it or looking over it) all vehicles with a windscreen are required to be fitted with one or more efficient automatic windscreen wiper(s), which must be capable of clearing the windscreen so that the driver has an adequate view of the road in front of the near and off sides of the vehicle in addition to an adequate view to the front.

C&U 25

1.2 Windscreen wiper(s) as required in 1.1 above shall at all times be in good and efficient working order and be properly adjusted. *C&U 95(2)*

2 Windscreen Washers

1 Every motor vehicle fitted with windscreen wiper(s) is required to be fitted with a windscreen washer capable of clearing, in conjunction with the windscreen wiper(s), the area of windscreen swept by the wiper(s). *C&U 26(1)*

3 Exhaust System

1 Every vehicle propelled by an internal combustion engine shall be fitted with a silencer, expansion chamber or other contrivance suitable and sufficient for reducing as far as may be reasonable the noise caused by the escape of the exhaust gases from the engine. *C&U 28*

2 All parts of a vehicle's exhaust system must be maintained in good and efficient working order. *C&U 31 & 98*

4 Audible Warning Instrument

1 Every motor vehicle must be fitted with an instrument capable of giving audible and sufficient warning of its approach or position. *C&U 27(1)*

Appendix E
Definition of a 'Dual-Purpose Vehicle'

"Dual-Purpose Vehicle" means a vehicle constructed or adapted for the carriage both of passengers and of goods or burden of any description, being a vehicle of which the unladen weight does not exceed 2040 kg and which either:

(1) Is so constructed or adapted that the driving power of the engine is, or by the appropriate use of controls of the vehicle can be, transmitted to all the wheels of the vehicle, or

(2) Satisfies the following conditions as to construction, namely:

(a) The vehicle must be permanently fitted with a rigid roof, with or without a sliding panel;

(b) The area of the vehicle to the rear of the driver's seat must:

(i) Be permanently fitted with at least one row of transverse seats (fixed or folding) for two or more passengers and those seats must be properly sprung or cushioned and provided with upholstered backrests, attached either to the seats or to a side or the floor of the vehicle; and

(ii) Be lit on each side and at the rear by a window or windows of glass or other transparent material having an area or aggregate area of not less than 1850 square centimetres on each side and not less than 770 square centimetres at the rear;

(c) The distance between the rearmost part of the steering wheel and the backrests of the row of transverse seats satisfying the requirements specified in head (i) of subparagraph (b) (or, if there is more than one such row of seats, the distance between the rearmost part of the steering wheel and the backrests of the rearmost such row) must, when the seats are ready for use, be not less than one thrid of the distance between the rearmost part of the steering wheel and the rearmost part of the floor of the vehicle.

C&U 3

Printed in the United Kingdom by Her Majesty's Stationery Office
at HMSO, Edinburgh Press
Dd 240200 C55 8/88 (256775)